Creativity as an Exact Science
The Theory of the Solution of Inventive Problems

Studies in Cybernetics

A series of books and monographs covering all aspects of cybernetics

Edited by F.H. GEORGE, Brunel University

(continued at back of volume)

Creativity as an Exact Science
The Theory of the Solution of Inventive Problems

G. S. ALTSHULLER

Translated from the Russian
by
Anthony Williams

Gordon and Breach Publishers
Australia China France Germany India Japan Luxembourg
Malaysia The Netherlands Russia Singapore Switzerland
Thailand United Kingdom United States

First published 1984
Third printing 1995

3 Boulevard Royal
L-2449 Luxembourg

Library of Congress Cataloging-in-Publication Data
Al'tshuller, G. S. (Genrikh Saulovich)
    Creativity as an exact science.

    (Studies in cybernetics, ISSN 0275-5807; v.5)
    Bibliography: p.
    Includes index.
    1. Creative ability in technology. I. Title.
II. Series.
T49.5.A47 1983   608'.01'9   83-16465
ISBN 0-677-21230-5

Contents

5 Forty Basic Methods

6 From Simple to Complex Methods

7 The Strategy of Invention: Controlling the Presentation of Problems

8 The Science of Invention

Introduction to the Series

The subject of cybernetics is quickly growing and there now exists a vast amount of information on all aspects of this broad-based set of disciplines. The phrase "set of disciplines" is intended to imply that cybernetics and all the approaches to artificial (or machine) intelligence have a near identical view-point. Furthermore, systems analysis, systems theory and operational research often have a great deal in common with (and are in fact not always discernibly different from) what is meant by cybernetics, as far as this series is concerned: inevitably, computer science is also bound to be involved.

The fields of application are virtually unlimited and applications are discovered in the investigation or modelling of any complex system. The most obvious applications have been in the construction of artificially intelligent systems, the brain and nervous system, and socio-economic systems. This can be achieved through either simulation (copying as exactly as possible) or synthesis (achieving the same or better end result by any means whatsoever).

The range of applications today has become so broad it now includes such subjects as aesthetics, history and architecture. Modelling can be carried out by computer programs, special purpose models (analog, mathematical, statistical etc.), and automata of various kinds, including neural nets and TOTES. All that is required of the system to be studied is that it be complex, dynamic and capable of "learning", and also have feedback or feedforward or both.

This is an international series.

FRANK GEORGE

Preface

It is not easy today to astonish anyone by the idea of controlling a particular process. Controlling thermonuclear energy? What is the problem? It is only a matter of the next few years before this is controlled. Controlling inheritance? How can you even ask? Haven't we genetic engineering already? Controlling the weather? Can there be any doubt? We shall certainly be able to order rain. Controlling the movement of the stars? Granted it is not an easy problem, but in principle there are no obstacles and, we shall learn how to control the stars, it is only a question of time. Any idea of controlling something that is not yet controlled today is received calmly. We will find means of controlling it, we shall control it. Yet only the idea of controlling the process of creativity as a rule encounters hefty resistance.

"Everyone knows that the act of creativity is not arbitrary", writes the playwright, Rosoy. "It does not respond even to the mightiest effort of will or peremptory command. However paradoxical it might seem, at the moment of creation the artist does not think as it were, since thought would kill creativity... It seems to me that the artist thinks right up to the moment of creation and also afterwards, but at the time of the act of creation itself he does not consciously reflect on it. Things are naturally more complex in scientific creativity. But it too is kin to the artistic — possibly even a blood sister. A few years ago I read a note in an article of the effect that the starting point of the greatest achievements and discoveries in all spheres of culture, science, technology and art is the sudden moment enlightenment which occurs unexpected and without evident cause. This is what creativity is" (*Question of Philosophy*, 1975, No 8, p.151).

I first came across this view of creativity 30 years ago when I began to study the science of invention. When talking about their work, scientists and inventors spoke with a striking unanimity about sudden enlightenment, the impossibility not only of controlling the creative process, but also of understanding what it is and how it comes about. Although people who had achieved a great deal in science and technology talked of the inscrutibility of creativity, I was not convinced and disbelieved them immediately without argument. Why should everything but creativity be open to scrutiny? What kind of process can this be which unlike all others is not subject to control? Many inventions have come too late, — this has been known for a long time. Inventors make frequent mistakes, dreaming up "legged" locomotives and sewing machines "with hands", so does this mean that this must always be so? I was determined to study this problem and assumed that in a year or two I would have solved it.

The problem turned out to be considerably more complex. Imagine being given the problem of making a sailing fleet independent of the wind. However, by their very nature sailing ships are wind-dependent, and there is nothing that can be done about it. Something similar emerged with inventive creativity. Problems of invention had been solved for a long time by a selection of variants. (And what if we do it so...?) This process also turns out to be dependent on a multitude of fortuitous factors which are difficult to comprehend, i.e. in practice and in fact invention problems were not subject to control. It was necessary to switch to another technology, which produces the same product — inventions — but in another process of production which is controllable, well organized, and effective. In short, if you want to be independent of the wind, build a steamship and give up your belief that there is no getting round sailing ships, even though all around you can see only sailing ships which indeed constitute the entire fleet.

Building a theory of the solution of inventive tasks even in outline is highly labour intensive. In the 1960s, a team of researchers was put together; the first public institutes and schools sprang up where one could test and polish the new technology of the solution of inventive problems. Today, in the 1980s, about 100 of these institutes and schools are at work. Each year the theoretical principles of the solution of inventive problems are studied by scientific workers, engineers and students; the volume of "production" is increasing rapidly since the graduates continue to invent after their studies are over. The theory is being improved, experience is being accumulated and this finds its reflection in the output of "production".

Before you is a book which descibes the new technology of creativity, in which the process of thinking is not chaotic but organized and plainly subject to control. This book can be read in two ways. One can simply read it straight through without going into too much detail — much as we would read a book about space flights and descends into the sea: they are interesting enough, but we would not ourselves want to fly to Mars or go down to the Marianna Trench. The reader retains the main thing in his memory: that there is a new technology of creativity, that if at some time a reader has to solve an inventive problem, he should begin not with a blind selection of variants, but by mastering the theory. One can read a book in another way — by working on it. Committing to memory the basic principles and rules, solving or at the very least trying to solve the problems included at the end of each chapter, and re-reading the chapter if the problems do not work out.

A word about the problems themselves. Naturally one cannot speak of the theory of the solution of inventive problems without presenting specimen problems. For that reason there are many of them in the book. Do not be afraid of them, do not leave them out if they are "not in your field". They are exercises in controlled thought, in overcoming psychological inertia, in applying the laws of the development of technical systems presented in the book. The solution of such problems needs no narrowly specialized knowledge, and what has remained in the memory from school physics will suffice.

The book is aimed first of all at the engineer. But it is also comprehensible to people who do not work with technology. The principles of controlling thinking in the solution of inventive problems (the principles and not concrete formulae and rules) can be transposed to the organization of creative thinking in any sphere of human activity. Therefore, the book is suitable for a wide readership. I hope that among those who read it there will be some who want to take it further and engage themselves in the search for new forms of the control of creative thinking in technology, science, and art. What can be more alluring than the discovery of the nature of talented thought and converting this thinking from occasional and fleeting flashes into a powerful and controllable fire of knowledge.

G.S. ALTSHULLER

1

On the Road to a Theory of Creativity

TRIAL AND ERROR METHOD

Invention is man's oldest occupation. The process by
which our remote forefathers became human began with the
invention of working tools. The first inventions were not
actually created by man, but were provided for him
ready-made. People noticed that with sharp stones it was
possible to cut the skins of dead animals - and they began
thereupon to collect and use stones. After forest fires it
was noticed that fire acts as a provider of warmth and
defence, and they began to keep fires alight. People had
not yet started to identify problems as such, but they
merely discovered ready-made solutions. The creative part
consisted in guessing how to apply these solutions. But
almost instantaneously problems of invention also presented
themselves. How to sharpen a blunted stone? How to make a
stone fit more comfortably in the hand? How to shield a
fire from wind and rain? How to transport fire from place
to place?

They had to resort to the trial and error method to
solve these problems of invention, and work through all
possible variants. For a long time the selection of
variants proceeded at random. But gradually certain
methods came to the fore. Copying natural prototypes,
increasing the size and number of objects acting in

1

conjunction, bringing together different objects into a single system. Facts, observations and information about the properties of substances were accumulated. Utilization of this knowledge enhanced the ability to direct the search and put some order into the problem-solving process. But the tasks themselves were also undergoing change and as century succeeded century they grew in complexity. Today, in order to find the one correct variant of a solution required it may be necessary to work through a host of "empty" trials.

Some traditional but incorrect views exist on the inventor's art. "It is all down to chance", say some. "Everything depends on diligence, you have to be keep on trying out different variants," maintain others. "It all depends on innate abilities" declare yet others. There is an element of truth, but only an external superficial truth in all these views. The method of trial and error is inefficient per se and therefore much depends on the problems and personal qualities of the inventor: not every one is capable of attempting "outrageous" trials, nor yet of applying himself to a difficult task and solving it with patience.

At the end of the 19th century Edison was using the trial and error method. His workshops employed up to a thousand men, and this made it possible to break down one technical problem into several tasks, and for each to simultaneously test many variants. Edison invented the scientific research institute (and this in our view is the greatest of his inventions).

Clearly a thousand navvies can dig more holes than one on his own. But their method of digging remains basically the same.

The modern "inventions industry" is organized along Edison's lines. The more difficult the task, the greater the number of trials needed to be carried out, and the greater the number of people needed to be employed solving it. Edison was able to give a group of three to five men the task of tackling "How to make the most reliable bond of a glass and metal part." Nowadays tasks of this order are tackled simultaneously by many teams, each consisting of dozens or even hundreds of scientists and engineers.

It is axiomatic that nowadays major inventions are made not by isolated individuals but by teams. Like every such belief it reflects only a partial truth. What is of paramount importance is not whether the work is performed by individuals or teams but the level of organization of work. The "lone" excavator driver works far more productively than a "team" of navvies. And a "team" of navvies can only loosely be considered a team - each digger works as an individual...

The trial and error method and the organization of creative labour based upon it stands in contradiction to the demands of the modern scientific revolution.

There is a need for new methods for managing the creative process capable of radically reducing the number of "empty" trials. Also needed is a new organization of the creative process which would permit the effective application of new methods. And this requires a scientifically based theory for the solution of inventive tasks capable of being implemented in work practice.

FROM THE HISTORY OF THE STUDY OF INVENTIVE CREATIVITY

In Volume 7 of his "Mathematical Anthology" the Greek mathematician Pappos who lived around 300 A.D. first coined the term "heuristics". And although Pappos cited Euclid, Appolonius of Pergamos and Aristos the Elder as his precursors, the origin of heuristics - the science of making discoveries and inventions - is associated with his name(1).

Later many mathematicians including Descartes, Leibniz and Poincaré applied themselves to the problem of creating heuristics. Evidently since it was denied the possibility of experimental development, mathematics felt the need for an instrument for solving creative tasks earlier and more strongly than other sciences.

From the first the terms "discovery" and "invention" were given the broadest interpretation in heuristics. Discoverers and inventors were taken to include artists, poets, politicians, military leaders, philosophers and others. In studying the techniques of mathematical creativity, mathematicians turned to factual material. They examined the process of solving mathematical problems, analyzed the experience gained during their study, and experimented with students. But as soon as they made attempts to formulate general laws of creativity these scholars broke away from the scientific approach and began to operate on uncoordinated facts, historical anecdotes, etc. Typical in this regard are the books of D. Poy (2) and G. Adamar (3) whose analysis, concrete and profound when dealing with mathematics, becomes superficial when it turns to general or technological creativity.

In Russia heuristics has been studied a great deal by the engineer P.K Engelmeier, the author of a number of works on the theory of creativity. "Heurology", he wrote, "is what I call the general theory of creativity, that is to say, the theory which embraces all phenomena of creativity, both artistic creation, technical invention, scientific discovery and also practical activity aimed at being used for anything at all. In this way wave theory is also heurology." (4. p.132). Engelmeier's books contain interesting materials, express a multitude of valuable ideas, in particular on the possibility of creating bionics. Engelmeier wrote "... it would appear that genius is not at all a divine and rare gift ... but is the destiny of everyone who has not been born a complete idiot."(4. p.135). Half a century later this notion was repeated word for word by Zwicky, the author of morphological analysis.

In the second half of the 19th century research studies on the psychology of scientific and technological creativity began to appear. Essentially these were also heuristics but with the accent placed on the psychology of thought.

At first scientific research was directed predominantly toward studying the personality of the inventor. In this period the creative personality was

regarded   as something exclusive.   There was much discussion
of   the   resemblance   between   psychological   disorders   and
genius,   about   the   special   constitution   of   the   blood   of
inventors,   etc.   And   only   in   the 20th century did these
views   gradually   give   way   to the conviction that creative
talents can be found in practically everyone.

Psychologists   started to experiment with simple tasks.
Especially   interesting   work   was   done by K. Dunker and L.
Sekey   (5).   It   emerged that the experimenters solve tasks
by   compiling   a list of variants, that in this process much
depends   on   their previous experience and that each variant
considered   transforms their view of the task, etc. However,
this   did   not   shed any light on the main problem of why do
certain   inventors   manage   to   solve a problem with a small
number   of   trials   when   it is generally thought   that they
will need   demand a larger number of trials?

The   psychology   of   creativity   cannot   answer   this
question   to   this day. In essence since the 1930's and 40's
no new results in principle have been obtained.

Why   should   psychologists   stubbornly   experiment with
simple   tasks and brain teasers instead of studying the real
process   of   creativity   involved   in solving complex tasks?
The   psychologist   N.P. Linkovsky (6) has rightly noted that
such   research   comes   up against practically insurmountable
hurdles.   The   creative process is extended over time. When
he   begins   his   observations   the researcher cannot be sure
that   the   "experimental   inventor" will manage to solve his
problem   even   in   the next five or ten years. And the very
observation   itself violates the sanctity of the experiment:

the greater the detail in which the psychologist questions the inventor, the more he learns of the course of his thinking, but also the more his questions affect the actual course of thinking itself, changing and distorting it. Although the creative process lasts for a very long time, the solution itself may appear suddenly, often in the form of a flash of "enlightenment". In this case it is simply impossible to ask questions. And in any case the facts as related by the inventor may not reflect the genuine course of thought. Back in the 1920s the philosopher I.I. Lapshin wrote: "It is extremely curious to note the deliberate striving of talented scientists who possess a deep knowledge of their subject and who are endowed with great sensitivity and perspicacity, to claim in front of laymen that their talent is some kind of mystic heaven-sent institution.(7, vol 2, pp 125-126).

For many psychologists even to this day the idea of guiding creativity sounds no more real than the idea of guiding the motion of the stars. At best it belongs in the distant future, and perhaps is forever unattainable. So the psychologists prefer to push their studies of creativity to one side and confine themselves to experiments with brain-teasers or simple chess challenges.

The theory of chess was created as a result of the compilation and analysis of a great number of complicated real matches. This approach is also possible in studying the creativity of invention. First of all one should amass and study a great number of descriptions of inventions. But whereas records of chess games may in some measure reflect the process of thinking of the chess players, the descriptions of inventions fix no more than the results of their work. One has to reconstruct the inventor's process of thought and for this one must oneself be able to solve difficult tasks covering various fields of technology.

At the basis of chess analysis lies the urge to understand how the game of Grand Masters differs from that of ordinary players. Sometimes a Grand Master can only be understood by a player who is equal to him in standing. A psychologist who takes the risk of entering the deep waters of studying the solution of complex inventive tasks has himself to solve tasks on a high level. This is difficult and so psychologists have to try to understand the inventor's creativity without themselves solving his problems. Only rarely in their experiments are tasks attempted similar to those of inventors. But then the attention of the researcher is concentrated only on the psychological factors. However, the psychological factors are secondary and conditional. The main thing in invention is that the technological system is transferred from one state to another, and according to definite laws and not "just at random". But precisely this is the primary and objective side of creativity which remains beyond the horizon of psychologists.

Imagine that we are studying the actions of a helmsman aboard ship on a meandering river. We want to know nothing about the river itself but only try to explain the actions of the helmsman in purely psychological terms. We see the helmsman beginning to spin the wheel rapidly to the right. Why? Maybe the sun is in his eyes and he is turning away from it - that is the reason... But now he is slowly turning the wheel to the left. Why? Perhaps he has decided to turn his face to the sun after all and get a tan? But now the helmsmen have changed over and the new man at once begins to turn the wheel and - Great God. - he has turned his back to the sun. All right, this means that the behaviour of helmsmen is determined by whether or not they like sunbathing, so let's note it down...

Unfortunately this is no exaggeration: this is precisely how the purely psychological approach, ignoring the existence of objective laws of development of technological systems, looks. In one of the chapters to come we shall be examining in detail an experiment conducted by Dunker and considered to be a classic. But first we shall be learning about the laws of development of technological systems and judging what is behind this love of sunburn...

Each science passes through the "alchemy" and "chemistry" stages. In the "alchemy" stage it tries to embrace the whole variety of the world in one of two formulae. Alchemy, for instance, looked on the field now studied by chemistry as something of tertiary peripheral importance. The alchemists were striving to obtain the philosopher's stone which would giving them eternal health, everlasting youth, wisdom, enable them to bringing the dead back to life and turn base metal into gold... The psychologists of creative thought are still on the "alchemy" level: they try by simple experiments to encompass the mechanics of any creativity. Building a general theory of creativity ought to precede the study of concrete aspects of creativity. Only by calling on the theories of inventive art, scientific creativity and literary creativity can one with time create a general theory of creativity which would in turn give a new push to the development of individual theories.

The road toward a scientific theory is long and difficult. However, real life practice and production demand new methods of solving creative tasks which are even marginally more effective than the simple compilation of variants. And such methods have appeared. These were purely psychological methods, but they were not created by psychologists.

METHODS OF ACTIVATING THE SEARCH

The more difficult the task of invention, the greater the variants which have to be worked through in order to produce a solution. And this being so the most urgent task is to raise the quantity of variants put forward in a given period of time. It is also understandable that if one is to discover a powerful solution the ideas inspected must include some which are more original, bold and unexpected ones.

The aim of methods of activating the search also consists in 1) intensifying the process of generation of ideas, and 2) raising the "concentration" of original ideas in the overall stream.

In solving a problem, an inventor at first spends a long time selecting these customary, traditional variants, which are close to his speciality. Sometimes he does not succeed in getting beyond these variants. The ideas point along the "vector of psychological inertia" toward where strong solutions are least to be expected. Psychological inertia is determined by the most diverse factors whether it be fear of entering an alien field, apprehension at putting forward an idea which might seem funny to some and ignorance of the elementary methods of generating "outrageous" ideas. Methods of activating the search help overcome these barriers.

The greatest renown among these methods is enjoyed by the brain storming put forward by A. Osborne (USA) in the 1940's. Osborne noticed that while some people are more inclined to generate ideas, others tend more toward critical analysis. In the usual discussions the fantasists and critics together in one place hindered each other. Osborne suggested separating the stages of generation and analysis of ideas. For 20 to 30 minutes the "idea generators" would launch several dozen ideas. The main rule was that criticism was forbidden. One could put forward any ideas including deliberately unrealistic ones, (which played the role of a kind of catalyst stimulating the appearance of new ideas). It was desirable that participants in the "storm" picked up and developed the ideas put forward.

If the "storm" was well organized they can quickly succeed in getting away from ideas imposed by psychological inertia. No one was afraid of putting out a bold idea and a benevolent creative atmosphere was created, which opened the way to all kinds of obscure ideas and conjectures. Usually taking part in the storm were people of different professions; ideas from varied fields of technology collided and sometimes this gave rise to interesting combinations.

The basic concept of the brain storm (to give new

ideas a way out of the subconscious) is founded on a theory of Freud highly popular in Osborne's country. According to this theory the "directed consciousness" is only a fine layer superimposed on the "undirected subconscious" like a hardened skin around the molten core of a volcano. Logic and control rule in the conscious preventing elementary forces, instincts, urges and desires from breaking out of the subconscious. In the conscious, order and clarity hold sway while down in the subconscious surge chaos havoc, stormy forces breaking through once in a while and forcing a man to commit illogical acts, to commit crimes, etc. Psychological inertia, in Osborne's opinion, is generated by order ruling in the conscious. One must help new ideas to break through from the subconscious to the conscious - such is the philosophical and psychological concept of the brain storm. For this reason Osborne constructed the process of generation of ideas in such a way as to unfetter the subconscious in a group of "idea generators" - who should not be the bosses - and one should strive to create a relaxed atmosphere. Sometimes appended to the storm concept was a kind of "rush" approach and the "idea generators" expressed proposals without having time to think them over. The ideas popped up seemingly willy-nilly, unthought-out and unguided. All the while the tape recorder registered every word. The ideas produced in this storm session were turned over to a group of "critics" for expert assessment. The critics' job now was to extract the rational kernel of each idea.

The following point is curious: In order to minimise the regimentation of thought (that kind of regimentation by which thought was determined by psychological inertia) one had to step up the orderly procedure of thinking and introduce certain ground rules. I wonder whether Osborne saw the paradox here?

Great hopes were placed on the brain storm in the 1950's. However it emerged that really difficult tasks did not lend themselves to this approach. Various modifications were tried out (individual, in pairs, mass, two-stage, the "conference of ideas" the "cybernetic session" and so on). These experiments still continue today. But it already clear that the brainstorm is effective only for the solution of simple tasks. Good results are obtained most often by applying storm techniques not to inventive but to organizational problems, (finding new applications for a product which is already being produced, improving advertising, etc).

Other methods exist for activating the search. For instance, the focal object method consists in transposing features of a few objects chosen at random to an object needing improvement as a result of which one can come up with unusual combinations and overcome psychological inertia. Thus if a "tiger" is taken as an accidental object and "pencil" as the (focal) object to be improved, then one obtains a combination such as "striped pencil", "rapacious pencil", "fanged pencil". By examining these combinations and developing them one can sometimes come up with original ideas.

In the morphological analysis put forward by the Swiss astrophysicist Zwicki the axes- the main characteristics of the object are first separated out, and then, for each axis the elements are noted, every possible kind of variant. For instance, in examining the problem of starting a car engine in wintery conditions one can take as the axes the sources of energy for heating, methods of transferring energy from a source to the motor, methods of controlling this transfer, etc. The elements for the "sources of energy" axis can be the battery, a chemical heat generator, a petrol heater, another engine driving the motor, boiling water, steam etc. By noting the elements along all axes and combining these different elements one can arrive at a large total of all possible variants. In this process one can also turn up unexpected combinations which would hardly suggest themselves of their own volition.

By the method of control questions, as the name suggests, the search is directed by a series of guided questions. Such series have been put forward by various authors. Typical questions are: What if you do the opposite? What if you replace this problem by another? What if you change the shape of the object? What if you take another material?

The most powerful method of activating the search is the synectics proposed by W. Gordon. In 1960 he set up the "Synectics" firm in the USA. The basis of synectics is formed by the brain storm, but this storm is conducted by a professional or semi-professional group which accumulates the experience of problem-solving gained from storm to storm. The synectic storm permits elements of criticism and, most importantly, it provides for the mandatory utilization of four special methods, based on analogy: direct (how are problems solved which are similar to the one in question); personal (try to put yourself into the shape of the object in the task and reason from that viewpoint); symbolic (give in two words a model definition of the essence of the task); and fantastic (how would figures in fairytales solve the problem).

The Synectics firm cooperates with the largest industrial firms, corporations and institutes of higher education in teaching synectic storms to engineers and students.

The main merit of these methods of activating the search is their simplicity and accessibility. Methods such as the brain storm can be mastered in one or two sessions. Studying synectics usually takes a few weeks in all.

Methods of activating the search are universal and can be applied for the solution of any tasks - scientific, technical, organizational etc.

The principal shortcoming of these methods is their unsuitability for solving rather difficult tasks. The storm (simple or synectic) throws up more ideas than the usual trial and error method. But this is not much if the "cost" of the task would be 10,000 or 100,000 trials.

Methods of activating the search retain (albeit in a rather improved form) the old tactics of selecting variants. These methods are not developed and attempts to combine them do not bring substantially new results. Therefore, in the Soviet Union these methods of activating the search have not found wide application.

LEVELS OF PROBLEMS

If you were to ask someone how to hunt, you would immediately be asked in return what you wanted to hunt. Microbes, mosquitoes, whales are all living creatures which can be hunted. But hunting for microbes, mosquitoes and whales takes on three qualitatively different forms. No one would study these three forms of hunt "in general." In invention, however, for a long time creativity was studied "in general" and conclusions drawn from "microbe" inventions were widely applied to "whale" inventions, and vice versa.

The scientific approach to the study of the inventor's art begins with grasping the simple truth that problems are various and can never be studied "in toto". There are some very simple problems which are susceptible to solution after a few trials, and others of unimaginable complexity which take many years to solve. What makes easy problems easy? Why are difficult ones difficult? What is it precisely that makes a problem difficult? Is it not possible find some means of transforming a difficult problem into an easy one?

Before examining these questions, let us first define the concept of "easy" and "difficult" problems.

One can divide problems according to their degree of difficulty into five levels or classes. The very easiest problems, (the first level) are characterized by the use of means (devices, methods, substances) intended precisely for the given goal. Here is an example of a problem of the first level.

PROBLEM 1

A furnace contains molten metal; into the central core of the furnace leads a pipe carrying liquid oxygen. What must be done to prevent the oxygen travelling through the pipe from gasifying until the point of exit from the pipe?

The answer is obvious: what is needed is heat insulation, and if it is already present it must be strengthened, thickened, double walls have to be added, enforced cooling, etc, has to be resorted to. This was precisely the way in which the actual problem was tackled: "A device for passing liquid oxygen into molten metal in the form of four concentric cooling pipes, and a muzzle designed to prevent the gasification of the oxygen flowing through the innermost pipe, insulated from the surrounding heat by an insulation layer 15 to 20 mm thick" (the author's description. Patent Certificate. No. 317 707).

To counteract heat you have to apply insulation of a thickness more than 1,5 to 2mm, which would be clearly too little, and not as much as 1,5-2 meters, since a tube with such a protective layer would take up too much space in a furnace. It should be 15-22 mm, as one would expect. The solution is patently obvious. Numerous experiments with the problem have shown that anyone can solve it with a few trials - scientific researchers, constructional designers, students, students at colleges of technology, schoolchildren. It is curious to note that Pat. Cert No

317 707 is attributed to ten authors.

This is a typical problem, solved at the first level: in principle the very same problem can also be solved at various levels.

In every edition of the <u>Bulletin of Discoveries</u>. <u>Inventions. Industrial Prototypes</u>. Trademarks around 30 percent of the inventions are solutions to such problems. In the given instance the search for a solution was practically reduced to zero. The technology of inventive creativity on this level has no need of improvement.

Let us now suppose that the problem is: "The arc light prevents the electric welder from seeing the processes going on in the welding zone. The light from the arc blanks out the duller parts - drops of metal, etc). What can be done?" Formulated like this the problem can be tackled without difficulty on the first level. The welding zone should be illuminated by a beam of light brighter than the arc itself. Now let us increase the difficulty of the problem by introducing supplementary specifications.

PROBLEM 2

The arc prevents the electric welder from seeing the processes taking place in the welding zone. The light from the arc blanks out the duller parts (drops of metal, etc). The conditions for observation must be improved without substantially complicating the apparatus and without lowering productivity.

This new problem is more complicated and therefore one

has to choose a few dozen variants. One rejects, for instance, all proposals involving introducing supplementary lamps for illuminating the welding area, since they would considerably complicate the equipment needed. Also inappropriate are all suggestions calling for periodically switching off the welding arc since this would involve reduction of productivity. The simplest solution, meeting the conditions laid down in the problem appears thus: "A device for shielding the eyes and face of the electro-welder consisting of a framework and frame with an inset light filter designed to improve observation of the process of welding; it is equipped with a reflector in the form of a right-angled section of a sphere with the same dimensions as the shield and focussing the light from the arc onto the welding material and the molten zone"(Pat. Cert No 252 549).

In problems of the first level the object (device or method) does not change (the heat insulation already present is strengthened). At the second level the object is changed but not substantially (a mirror is introduced into the shielding device). At the third level the object is changed essentially and at the fourth it is totally changed; in the fifth the entire technical system is changed in which the object fits.

An example of an invention of the third kind: "A nut and bolt combination designed to prevent their surfaces wearing out by eliminating friction between them during their movement; a clearance is allowed between them which is retained while they are operating and into the threads is laid a coil setting up an electromagnetic field

guaranteeing a progressive motion of the screw relative to
the bolt" (Pat. Cert No 154 459). The nut and bolt remain
but they have been radically altered in relation to their
prototype.

As an example of inventions of the fourth type one
could take a new method of inspecting engine wear.
Previously inspection of wear was carried out by
periodically taking an oil specimen and measuring the
number of metal particles it contained. According to Pat.
Cert. No 260 249 it is proposed to introduce into the oil
luminescent traces and by monitoring the changes in the
amount of illumination (tiny particles of metal extinguish
radiation) maintaining a constant check on the
concentration of metal particles. The original method has
been replaced completely. The physical effect utilized here
is less well familiar than in the previous example. The
idea found has broader application than the patented method
of checking wear: from the extinction of luminescence one
can monitor the appearance of metal particles in other
applications as well.

An invention of the fifth type: "The Application of
monocrystals of alloys of copper - aluminium - nickel and
copper -aluminium - manganese as a solid working body for
transforming heat into mechanical energy by the changes in
its resilient properties with variations in temperature.
But we know of few substances which alter their properties
radically with small changes in temperature. Finding or
obtaining such substances of itself borders on a discovery.
New transformer substances can be utilized for solving the
most varied problems of invention (making heat engines,
various measuring devices, etc.)"

Solution of a problem of the first sort requires choosing between a few obvious variants. This can be performed by any engineer and such problems are solved every day without difficulty, although they do not always find expression in patent applications. On the second level the number of variants is already measured in dozens. Any engineer is in principle capable of picking out 50 to 70 variants. But nevertheless this demands a certain patience and persistence and certainty of the possibility of the problems being solved. Sometimes a man gives up after only ten trials. A correct solution of problems of the third level may be buried among hundreds of incorrect ones. Thousands and tens of thousands of trials and errors can be made on the fourth level in order to find a solution. Finally, on the fifth level, the number of trials and errors grows to hundreds of thousands and millions. One recalls, for instance, that Edison had to make 50,000 trials to invent the accumulator battery. The question here is only about essential experiments; notional experiments, all possible "And what ifs?" would probably be considerably greater.

Here is an example of a teaching problem of the fourth level.

PROBLEM 3

Uneven planks and twigs of trees are reduced into chippings, consisting of a mixture of bark and wood chips. How can the pieces of bark be separated from the chips of wood if they differ very little in density and other aspects?

A multitude of patents have been issued in various countries to cover this problem. Inventors have stubbornly (and unsuccessfully) tried to separate pieces of bark from wood chips by capitalizing on miniscule differences in density. In experiments on this problem the number of trials have sometimes gone into the hundreds, yet none have succeeded in overcoming the psychological barriers and taking a principally new and, this is the main thing, true direction.

The question to be asked is, if they make inventions of a high level succeed does this still mean working through hundreds and thousands of variants?

Here a very interesting relay mechanism comes into effect. A problem has appeared and the 'price' is 100,000 trial attempts. Someone has spent half his life on working through 10,000 trials and not come up with the solution. Another person sets about solving the problem, he digs up yet another part of the field being searched, and so on. The problem acquires the reputation of being insoluble, of taking a hundred years. In fact, however, it is gradually being simplified until it is finally solved. At this point researchers come on the scene who try to explain the secret of the inventor who has solved an "eternal" problem. Yet there is no secret. The failures, those who assaulted the problem in relays could have been even more successful in the beginning than the one who ran the final stage. It was simply that they were given a too broad field of search. In essence, the problem is solved not by one man but a whole team, "a cooperation of contemporaries" in Marx's definition. For very complex problems it may even be necessary to have a cooperation of inventors covering several generations. Their efforts gradually turn the problem of the fifth level into a comparatively simple one of the first level, until someone takes the last hurdle by the very same method of trial and error.

There is another method, which can be called "the problem finds its own solver." A complex problem is difficult because it relates to one field, and for its solution knowledge from quite a different field is needed. When in 1898 Crookes presented the problem of binding nitrogen, thanks to his scientific reputation this became known to very many scientists. Birkeland, the Norwegian specialist in polar aureolas, proposed using the processes similar to those going on in the upper atmosphere. The problem "sought out" the man whose special knowledge was necessary for its solution.

The higher level problems differ from the lower not only in the number of trials it takes to find a solution. There is also a qualitative difference. The problems of the first level and the means of solving them are to be found within the confines of one narrow speciality (the problem of improving the production of chipboard is solved by methods also utilized within the industry). Problems of the second kind and the means of solving them relate to a single field of technology (the problem of the chipboard is solved by methods well-known in forestry). For solving problems of the third level, one has to turn to other fields (the problems of the timber industry are solved by

methods well-known in metallurgy). The solution of problems
of the fourth level must be sought not in technology but in
science,   usually   among   little utilized physical and
chemical effects and phenomena. At higher sub-levels means
of solving problems of the fifth level may in general turn
out to be beyond the limits of contemporary science.
Therefore first one must make the discovery and then,
relying on the new scientific data, solve the invention
problem.

On the first and second levels one can choose variants
using   the   knowledge   relating   to   one's particular
speciality. The higher the level, the broader the knowledge
called for. A good team of specialists can easily make
inventions of the first and second level. Such inventions
improve technology. But solutions which are radically new
in principle should be expected to come rather from
"outsiders". Take, for example, Pat. Cert. No. 210 662: "An
electromagnetic induction pump containing a chassis, an
inductor and a channel designed to simplify the action of
the pump; the inductor is filled with a lubricant along the
axis of the channel of the pump." This invention was made
by   specialists: no   revolution but a quite passable
improvement. The board of assessors easily approved the
new idea - from the date of application to publication
took only 14 months. But the journalist A. Presynaykov took
fourteen years to obtain the patent certificate (No 247
064): (The application of an electromagnetic pump for
pumping electrolytes as a nautical jet engine. At the
basis of this invention was a magneto-hydraulic effect. The
idea was put forward at a time when practically nothing was

known about the magneto-hydraulic engines which now enjoy such fame.

One further example: four students at Public Institute of Inventive Creativity took as the subject of their diploma dissertation a highly complicated problem in the field of aeronautical navigation. Work was being carried on in this field in many countries. Three of the students and one young engineer were not specialists in this field. They calculated that a powerful solution would be found outside the range of accepted ideas and principles of navigational instrumentation. So it turned out. The necessary principle of analytical measurements was found in the technology of the confectionery trade which is a long way removed from aviation. The invention received a positive evaluation from the assessors and a patent was awarded.

The scientific and technical revolution demands that problems of a higher order be solved in ever shorter periods of time. The usual method of intensifying the process of finding solutions is to increase the number of personnel working simultaneously on one and the same project. But the possibilities of such an intensification are limited. The concentration of large numbers on solving a single technological problem can lead to a diminution of the intensity of work in other directions.

What is needed is a method of transposing inventive problems from higher to lower levels. If one can succeed in translating a problem of the fourth or fifth level to the first or second, then you can make a successful job of selecting variants. The whole problem consists in being

able to rapidly narrow down the field of search, converting
a 'difficult' into an 'easy' problem.

CONTRADICTIONS, ADMINISTRATIVE, TECHNICAL AND PHYSICAL

Let us compare two inventions. The first, "A Method of
Defining the Parameters Inaccessible to Direct Observation
(For instance, resistance to wearing out) based on
indirect inspection designed to raise the accuracy of
determining the unknown parameters based on the results of
indirect inspection; manufactures are selected in pairs
(series) according to the principle of the proximity of
measured parameters within the same model from each pair
(series); the unknown parameter is determined by destroying
the manufacture and applying the result obtained to the
remaining wares in this pair (series)."(Pat. Cert.. No 188
097). In order to check the wares they propose an extremely
simple solution: smashing up half the number and taking a
look... True, a contradiction arises here in that the more
wares are smashed up the greater the reliability of our
judgement of those remaining.

The second invention: "A method of inspection and
fault detection of goods of the same type having hidden
defects, for example in the form of bubbles or foreign
matter, designed to simplify the process of controlling the
product by placing it in a vat containing an electrical
conducting liquid, passing through it an electrical current
and then bringing to bear on the liquid a magnetic field
for measuring the apparent density until a mean situation
is obtained where perfect products and those with defects
are determined by measuring their position vis a vis the

bottom of the vat. (Pat. Cert.. No 286 318). This problem
is very similar to the first but in its solution there is
no contradiction - the experiment is carried out without
destroying the product. An original method has been
employed using the interaction of electrical and magnetic
fields to force the liquid to as it were change its
density, as a result of which an object placed in it sinks
or rises (depending on the presence or absence of defects
in it.)

The inventor's problem is often confused with problems
of technology, engineering or design. Building a normal
house, having readymade blueprints and calculations is a
technical problem. Making the calculations for a normal
bridge, making using of available formulae is a matter for
engineers. Designing a comfortable and cheap bus, finding a
compromise between 'comfortable' and 'cheap' is a
designer's business. In solving these problems one does not
have to overcome any contradictions. The problem becomes
that of an inventor only in the event that contradictions
have to be overcome for its solution.

We do not come across contradictions even when solving
problems on the first level. Strictly speaking these are
designers' and not inventors' problems. In legal terms
'invention' does not coincide with what we would call the
technical or creative understanding of the word. Evidently
with time the juridical status of inventions will change
somewhat and simple designers' solutions like that
described in Pat. Cert. 317 707 (the introduction of heat
insulation) will cease to be considered inventions. To
avoid confusion we shall for the time being use the phrase

'inventive problem of the first level' remembering, however, that genuine inventive problems of the second and higher levels are by definition connected with overcoming contradictions.

In point of fact a contradiction is already present in the origin of inventive problems. Something has to be done, but how to do it is unknown. Such contradictions are customarily called 'administrative'(AC). There is no need to discover administrative contradictions since they lie at the surface of the problem. But the heuristic 'prompting' force of such contradictions amounts to nil. They do not say in what direction the solution should be sought.

Below the administrative level lie the technical contradictions (TC); if by certain methods one improves one part (or one parameter) of a technical system, it is inadmissible for another part (or another parameter) to deteriorate in the process. Technical contradictions are often indicated in the conditions stipulated in the problem, but just as frequently the original formulation of the TC requires serious correction. On the other hand a correctly formulated TC possesses a definite heuristic value. Admittedly the formulation of the TC does not give any pointer toward the answer in specific terms, but it enables one to throw out at once a multitude of "empty" variants: unsuitable by definition are all those variants in which a gain in one quality is accompanied by a loss in another.

Each TC has specific physical causes. Let us take as an example the following problem:

In polarising optical lenses it is necessary to put a cooling solution beneath the polishing surface (which is made of resin). An attempt was made to incorporate in the polishing machine slit-like orifices through which a liquid could be squirted at intervals, but the perforated surface of the polisher then performed worse then the polisher as a whole. What can be done?

The technical contradiction here has already been indicated: the cooling capability of the "perforated" polisher comes into conflict with its ability to polish the glass. What is the reason for the conflict? A "hole" is good at letting a coolant through, but, not unnaturally, is no good at abrasing a piece of glass. The solid sectors of the polisher on the contrary are capable of abrasing particles of glass but are unable to let water through. Hence the surface of the polisher has to be solid in order to abrase the particles of glass yet "holed" in order to let the coolant through. This is a physical contradiction (PC): mutually opposing demands are placed upon one and the same system.

In physical contradictions the conflict of demands is intensified to the maximum. Therefore at first glance the PC seems absurd, inadmissible by definition. What can one do in order that the whole surface of the polisher be one big "hole" and at the same time an entire solid body? But it is in precisely this, carrying the contradictions to the extreme, that the heuristic strength of the PC shows itself. Since one and the same substance cannot be in two different states it only remains to take apart, to disunite the contradictory properties by simple physical transformations. One can, for instance, divide them spatially: let the object consist of two parts, possessing different properties. One can divide the contradictory qualities in time: let the object take it in turns to have first one property then the other. One can utilize transitory states of the substance in which something akin to co-existence of the contradictory properties obtains temporarily. If, for instance, a polisher is made of ice with particles of the abrasive frozen into it, in polishing the ice would melt, ensuring the requisite combination of properties. The polishing surface would remain unbroken and at the same time cold water would pass through it at every point, as it were.

THE KEY TO THE PROBLEM: LAWS OF DEVELOPMENT OF TECHNICAL SYSTEMS

Thus methods are needed which permit one to discover and eliminate the physical contradictions inherent in inventive problems. These methods allow one to narrow down the field of search severely and, without "piecemeal" checking to reject the majority of "empty" variants.

We have already named a few methods: separating the contradictory properties in space or time, utilising transitory properties of substances. What else? Where can one take an assemblage of methods sufficiently rich in order to solve the most varied inventive problems? The answer is obvious: physical contradictions are inherent only to inventive problems of the highest levels, therefore methods of eliminating them should be sought in the solutions to these problems. In practice this means selecting inventions of the highest levels and studying descriptions of them. These descriptions usually indicate the original technical system, its shortcomings and the proposed technical system. By comparing this information it is possible to arrive at the essence of the PC and the method utilized for its elimination.

The storehouse of descriptions of inventions is extremely large. Each year in various countries around 300,000 patents and copyrights are issued. To arrive at contemporary methods of eliminating physical contradictions it is sufficient to study the latest "patent layer" to a depth of, say, five years - representing around 1.5 million inventions. The number is terrifying. However, the very first move is to go for inventions of the highest level - since this sharply reduces the number of descriptions needing detailed scrutiny. Inventions of the fifth level are very few - a fraction of one percent; those on the fourth level are also few - three to four percent. Even if you take the more interesting inventions of the third level you have to look at no more than 10 percent of the inventions in the "patent layer" exposed: 150,000 descriptions. That would be the maximum. In order to compile a list of the most powerful examples a batch of 20-30,000 patent descriptions would be sufficient.

A good list of methods of removing the PC alone would be a lot in itself. But one has to have correctly formulated contradictions and also to know when and what method to use; one must be in possession of criteria for evaluating the results one has obtained. For this it is necessary to know the rules of development of technical systems.

The development of technical systems, like all other systems is subject to the general laws of dialectics. In order to concretize these laws applicable precisely to technical systems once has to turn once again to the patent storehouse but going now to a considerably greater depth. One must take not the "patent layer" but a "patent borehole" so to speak - an extremely, highly complex procedure. But, knowing the rules of development of technical systems, one can confidently pick on the more effective methods for eliminating contradictions and put together a programme for solving invention problems.

What are the objective laws governing the development of technical systems? Let us take a specific example. A cinema studio is a typical technical system comprising a number of elements: a cine camera, lighting, sound recording equipment, etc. The camera shoots at a speed of 24 frames per second; for the exposure of each frame the shutter opens for a very short interval of time, sometimes as little as one thousandth of a second. Yet the lights operate on continuously current (or interrupted, but this possesses greater heat inertia) and provides illumination constantly for the studio scene. Thus it would be useful to use far less energy on them. Energy is usually wasted and in a harmful way: it exhausts the actors and heats up the surrounding air.

Please note: the basic elements in this system "live" each according to their own rhythm. Imagine an animal whose brain works on a 24-hour cycle, but whose legs prefer to be active, say, on a ten-hour cycle. When the time comes round for the brain to sleep the paws are still active , full of vigor. According to their "clock" it is noon, they need to run... Evolution has pitilessly rejected such organisms. But technology very often throws up "organisms with an uncoordinated rhythm" and then for a long time one is tormented by the defects inherent in them.

One of the objective rules of development of technical systems is that systems with uncoordinated rhythms are pushed out by better systems with more harmonized rhythms. Thus, in the example quoted, it is necessary to have non-inertial lighting working in synchronization and in phase with the rotation of the lens shutter. This cuts down severely on the expenditure of energy, and improves the working conditions for the actors.

Let us take the example from another field of technology. In order to extract coal we dig a borehole into the seam, fill it with water and transmit through it pressure impulses. The frequency of the impulses is determined by fortuitous factors, and the seam has its own frequency of vibration. Again both parts of the system work in different rhythms - a clear violation of the law of coordinating rhythms. And so appears Pat. Cert. Nr. 317 797 in which it is proposed that a frequency of impulses be established which is equal to the inherent frequency of vibration of the coal mass.

Inventions (simply 'impulses' or 'impulses with a frequency equal to the innate frequency of the mass being drilled') are divided into intervals of seven years. These seven lost years are the price paid for ignorance of the laws of the development of technical systems.

Synchronizing the rhythms of parts of the system is only one of the laws governing the development of technical systems. By calling on the "sum" of such laws it is possible to construct a program for solving inventive problems. This would make it possible, without blundering about the field of search to enter the region of solution, that is, to pare down the number of variants, let us say, to about ten.

Furthermore, it would seem to be quite simple to look at the ten variants and pick out the one you need. But ten variants obtained by translating problems to the first level can differ qualitatively from the ten variants needed for solution of problems which from the outset were on the first level. With 'natural' problems of the first level all variants of solution are comprehensible to the inventor, usually relating directly to his speciality, and do not frighten him by their complexity. An "artificial" first level problem taken from, say, the fourth level can have solutions which appear "outrageous" or which transcend the extent of knowledge of the inventor. Let us suppose that analysis has pared away all "empty" variants, leaving only one possible behind. "The problem can be solved if a liquid rotating in a vessel is pressed not to the walls of the vessel but to its axis." It is commonly known that a rotating liquid is subject to centrifugal forces pressing in the direction of the walls. It is most likely that an inventor would reject such a variant as clearly contrary to physics... However, there are liquids in which, contrary, to all normal concepts, with rotation centripetal forces are exerted! This phenomenon is called the Weissenberg Effect (8, page 140). It is outside the physics taught in colleges and universities for engineers and therefore not all engineers know of it.

For a guaranteed solution of problems one needs information about the whole of physics. Precisely everything because the solution of difficult problems often involves calling on little known nuances of usual physical effects. Moreover, the whole of physics should be presented in such a way that effects do not have to be presented one after another. In other words, what is needed is not simply physics, but tables linking the types of inventive problems (or the types of contradictions) to the respective physical effects. This is the very same form in which to present the purely inventive methods brought out analysis of patent materials.

But that alone is not enough. The inventor, working on a program should not be afraid of rejecting plausible variants or of taking up seemingly "outrageous" ideas. In other words, what is needed is control of psychological factors.

Thus an effective technology for the solution of inventive problems can be based only on a conscious adherence to the laws of development of technical systems;

Deriving from these laws, one can construct a program for solving inventive problems allowing one, without selecting all variants to reduce problems of higher levels, to those of the first order;

To overcome physical contradictions the program should have at its disposal a pool of information, including an

archive of inventive methods amassed after analysis of
large masses of contemporary patent information; the fund
of methods should be presented in the form of tables on the
utilization of methods according to the type of problem or
the contradictions inherent in it;

The program should have means of controlling psychological
factors, in first place to stimulate the imagination and of
overcome psychological inertia.

AN ALGORITHM FOR THE SOLUTION OF INVENTIVE PROBLEMS (ASIP)

A program which all these requirements has been called
ASIP (Algorithm for the Solution of Inventive Problems).

The word algorithm in the narrow sense denotes an
absolutely determined sequence of mathematical operations.
In the broad sense the word "algorithm" is any sufficiently
clear program of action. It is precisely in this sense that
ASIP is called an algorithm.

It is important, however, to emphasize that with each
new modification the main signs of the algorithm are
strengthened: the determinism, mass, results.

Outwardly ASIP represents a program for consistent
processing of inventive problems. The laws of development
of technical systems are lodged in the structure of the
program itself or come out in their "working clothes" in
the form of concrete operators. With the help of these
operators the inventor can step by step (without empty
trials) get to the PC and determine that part of the

technical system to which it "belongs". Then the operators which alter an individual part of the system and eliminate the PC. By this means a difficult problem (i.e. a problem not of the first level) can be transformed in an easy one (of the first level).

The ASIP possesses special means of overcoming psychological inertia. Certain authors have assumed that it is not difficult to cope with psychological inertia, and it is enough to be reminded of its existence (9. pp 38-39). If only this were so! Psychological inertia is incredibly strong! What is needed are not appeals to remember it, but concrete operators for changing the problem. For instance, the conditions of the problem should be mandatorily stripped of special terminology, because terms shackle the inventor to old and ingrained concepts about the object.

In working out ASIP a systematic analysis of the patent bank was undertaken. Inventions of the third and even higher levels were isolated out and studied to determine the technical and physical contradictions contained in them, and the typical methods of removing them. In order to build up the table of applications of typical methods in one of the latest modifications of the ASIP, around 40,000 descriptions of selected inventions of the higher level were analyzed. In the course of the next three years the table has been corrected: prognostic corrections have been introduced and checked in new and complex problems. Such a table not only reflects the collective experience of a huge number of inventions but also has a solid reservoir of prognostic durability. The methods recommended by it will not become obsolete in the next 10 to 15 years.

With new modifications of ASIP, tables have been worked out for the application of physical effects and a detailed handbook "The Index of Applications of Physical Effects and Phenomena" has been compiled. With the aid of these tables one can determine the effects most suitable for overcoming the contradictions in the problem. The "Index" gives information about the effects themselves and substances released by these effects.

In essence ASIP organizes the thinking of the inventor enabling him to call upon the experience of all (or very many) inventors. Most importantly, this experience is utilized in a talented way. The normal and even highly experienced inventor draws from his experience solutions based on external analogies. He looks at this new problem and sees in it a similarity to such an such an old one, and hence the solutions should be similar too . The "ASIP" inventor, however, looks far deeper: he sees that in this new problem there is such an such an PC, which means he can use the solution to an old problem, which outwardly is not at all similar but which contains an analogous PC. To the outside observer this seems to be a manifestation of a mighty intuition.

The information apparatus of ASIP is regularly topped up and improved. In general ASIP develops very quickly. Modifications of ASIP have indices denoting the year of publication, and not just the usual number. A clear reference to the "year of issue" makes a systematic improvement of ASIP mandatory, and does not allow it to grow obsolete.

FROM ASIP TO THE THEORY OF SOLUTION OF INVENTIVE TASKS

With the appearance of the first modifications of ASIP began the establishment of the Theory of the Solution of Inventive Problems(TSIP). The correlation between ASIP and the theory is approximately that of an airplane and aviation, between a car and automobile transport.

The theory is incorporated in ASIP although, of course, it does not amount to the same thing. In coming chapters we shall have to touch equally on the concrete mechanics of ASIP and on the general positions of theories and how they are mutually interconnected. A few words about terminology. They are not simple and hence we shall talk about their content.

Approach: single (elementary) operation. The approach can apply to the actions of a man solving a problem "by analogy." It can also apply to the technological system examined in the problem, for example, "splintering of the system", the "uniting of several systems in one". Approaches are "scalar", so to speak, i.e. undirected. It is not generally known when this or another approach is good or bad. In one case analogy can lead to the solution of the problem and in other away from it. Approaches do not develop as such (although on the contrary they can, of course, be augmented and expand.

Methodology: a system of operations providing for a certain order of their application. For instance the brainstorm method includes a number of operations while recruiting a group of "idea generators" and "critics" while conducting the storm, while assessing the ideas.

Methods are usually based on some single principle or postulate. Thus at the foundation of the brainstorm lies a supposition that the solution of a problem can be found, which will provide an exit from the subconscious for an undirected stream of ideas. Methods are developed in an extremely limited fashion, remaining within the framework of the original principles. This is the sense in which we shall utilize the word "methodology".

Theory - is a system of many approaches and methods providing for a goal-orientated direction of the process of problem-solving based on knowledge of the laws of development of the objective reality. Crudely expressed, approach, method and theory form a chain of the type "brick-house-city" or "cell-organ-organism". In this hierarchy ASIP is to be found on the borderline between method and theory.

Work on ASIP was begun in 1946 (10-19). The concept of ASIP did not then yet exist as such and the problem was presented differently: "One has to study the experience of inventive creativity and bring out the characteristic features of good solutions, distinguishing them from the bad ones. The conclusions can be used in solving inventive problems."

Almost immediately they began to discover that the solution of inventive problems turned out to be good (strong) if it overcame the technical contradiction contained in the problem presented to it, and bad (weak) if the TC was not revealed and eliminated.

Furthermore something completely unforeseen happened: it became clear that even the strongest inventors did not

see or understand that the correct tactic for the solution
of inventive problems ought to consist in step by step
elucidating the TC, studying its causes and removing them,
and by so doing also removing the TC itself. Having
discovered the TC which was crying out to be seen, and
seeing that the problem was successfully solved thanks to
its removal, inventors drew no conclusions for the future,
and did not change their tactics. When starting in on
their next problem they might spend years ploughing
through the variants, without making an effort to
formulate the contradiction contained in the problem.

     The hope that one could draw from the experience of
big (great, influential, experienced, talented) inventors
something useful for beginners proved to be misplaced.
Great inventors worked with the selfsame primitive method
of trial and error.

     The second stage of work began, the problem now
sounded like this: "A program should be drawn up for the
planned solution of inventive problems, applicable to all
inventors. This program should be founded on the
step-by-step analysis of the problem, in order to
facilitate study and elimination of the technical
contradiction. The program would not replace knowledge and
ability, but it would prevent many mistakes occurring
and would be a good tactical approach to the solution of
inventive problems."

     The program for inventive problems was still far away
from the ASIP of today, but with each new modification it
became clearer and more reliable, gradually acquiring the
characteristics of programs (precepts) of the algorithmic
type. The first tables were compiled for applying methods

of removing technical contradictions. The main material for research became the patent information, the descriptions of inventions. The first study seminars were held, gradually accumulating experience of studying ASIP.

And once more something disconcerting occurred. It became clear that the solution of problems of the higher levels calls for knowledge necessarily going beyond the boundaries of the inventor's speciality ; experience in production necessitates fruitless trials in the customary direction. The only 'ability' to tangibly affect the progress of a solution is the 'ability' to adhere to an ASIP and utilize its provision of information.

Hence the unavoidable conclusion is that neither knowledge, nor experience nor ability (a 'natural gift') can serve as a reliable basis for the effective organization of creative activity. There are no people able to regularly solve a string of problems of higher levels by their knowledge, experience and abilities. If the 'price' of a problem is 100,000 trials no one can solve it in isolation.

In tackling an inventive problem of the highest level, a man has to have a knowledge of the whole of technology, the whole of physics, the whole of chemistry. However, the sum total of everything a man knows is millions of times less than this. In solving a problem a man should be able to correctly reprocess the information he possesses (assuming it is available in its full volume). By 'correctly reprocess' we mean building a chain of consistent actions, and directing these in such a way that they lead to a solution of the problem. Instead of this man resorts to primitive selection of variants, guided only by old concepts and personal (and hence fortuitous) experience.

Man cannot effectively solve inventive problems of the highest levels. Therefore all hypotheses which directly or indirectly proceed from the premise that by studying the creative process they can discover effective approaches, methods, heuristics, etc, are erroneous. All methods and methodology which are based on the urge to activate creative thinking, are erroneous, since they are attempts to organise bad thinking well.

In this way the second stage which began with the idea that inventors need to be given a useful aid, ended with the conclusion that a reconstruction of inventive creativity, and changes in the very technology of production of invention were necessary.

The program has now come to be regarded as an autonomous system, independent of man, which can solve inventive problems. If thinking followed this system and let itself be guided by it - then it would be genuinely talented.

The need arose for operations produced in the algorithm for the solution of inventive problems to be put onto an objective basis, and for them to be founded on the objective laws of the development of technical systems.

The argument for the third stage ran like this: "Inventions on the lower levels are not creative at all. Inventions of the higher levels carried out by the trial and error method are mediocre. What is needed is a new technology for solving inventive problems, which would permit one to solve problems of higher levels according to plan. This technology should be based on knowledge of the objective laws of development of technical systems.,

Toward the beginning of the third stage the system of

Public Institutes of Inventive Creativity began to get established. In 1978 there were already 100 of these institutes (in Moscow, Leningrad, Baku, Volgograd, Gorkiy and other cities.) Developing the theory, experience and state of the art of ASIP, and organizing these studies became a collective task involving the active participation of many researchers. By their joint efforts they managed to build up the information service of ASIP and in particular to compile the "Index of the Application of Physical Effects and Phenomena". A start was made on the so-called S-Field analysis, which linked the process of solution of problems with certain fundamental laws on the development of technical systems and allowed one to map out the routes for the systematic discovery of physical effects necessary for the solution of problems.

As in the second stage, the basic material was supplied by patent information. But study of this has now led not so much to the discovery of new methods and their inclusion in the table of elimination of technical contradictions, as to the study of the general patterns of the development of technical systems. Knowledge of these patterns allowed us to introduce correctives into ASIP and S-Field analysis.

The system of schools and Institutes of Inventive Creativity enabled us to test out in practice, quickly and reliably, new conclusions, suppositions and hypotheses.

The third stage is still in effect. But already something new has been discovered which is leading to further progress in the ideal suppositions of theories and to the entry of theories into a fourth stage of development. It has become obvious that the main thing is that an invention is the development of a technical system. The problem is only one of the forms in which the demands of development of a technical system are revealed by man. With the help of theory one can develop technical systems according to plan, without waiting until the problems arise.

PROBLEMS

We have not yet examined the mechanisms of ASIP and so far we have at our disposal only the customary method of trial and error. Try to solve several problems by using a selection of variants. In future we shall return to these problems and see what can be done using ASIP and the theory of the solution of inventive problems. No specialized knowledge is required for the solution of these problem.

PROBLEM 5

Extract from a detective novel:

"I didn't kill him, Sheriff, you've got to believe me, you must believe me!"
"I've only got to believe the facts", retorts the Sheriff. "The facts are against you, fella. Just this week you threatened Bolton - we've got witnesses. Bolton was killed with a bullet from a Colt. Exactly the same kind of Colt as yours. We didn't find the bullets, that's true, but our expert says the calibre is the same. What's more, you ain't got an alibi!"
"You gotta believe me!", exclaims Nick in

despair. "I didn't shoot, I swear it. You can see for yourself, my gun is completely clean..."
The Sheriff smiled.
"The murder took place two full days ago," he said. "You have had time to clean your gun..."
Imagine that you have been called in as an expert. You have to solve the problem from the position of an inventor.

PROBLEM 6

At a factory turning out agricultural machinery there is a small piece of ground for testing machinery (such as ploughs) on their ability to move forward, turn, etc. However, the "manoeuvrability" of machinery depends on the state of the ground. The need has arisen for conducting tests on two hundred different types of soil. It is impossible to build two hundred different testing grounds. What can be done?

PROBLEM 7

It is necessary to measure from the air the depth of a river every 300 to 500 meters for a length of 100 kilometers. The aircraft has no special equipment, landing personnel from it has been ruled out, and the survey has to be carried out as cheaply as possible. The accuracy of measurement has to be plus or minus 0.5 meters. The speed of current is unknown. How can it be done?

PROBLEM 8

A metal cylinder is being polished from inside using an abrasive disk. In the process of work the disk wears away. How can one measure the diameter of the disk without stopping the polishing process and removing the disk from the "bowels" of the cylinder?

2

Principles of S-Field Analysis

The S-Field - the Minimal Technical System
Let us examine a few inventive problems.
PROBLEM 9

Some means is called for enabling rapid and accurate identification of faults in refrigeration equipment through which liquids (freon, oil, hydroammoniac solution) can seep away.

PROBLEM 10

How can one determine the degree of hardening of a polymer mass when making items out of polymers? It is impossible to measure directly (by 'feel').

PROBLEM 11

How can one inspect the intensity of movement of particles of a bulk material in a state of pseudo-liquefaction?

PROBLEM 12

Suggest an easily extractable wedge.

These problems are drawn from various fields of technology, and describe various situations each of which has its own pitfalls. Problem 9 calls for the rapid and accurate search for small droplets of a liquid whereby "rapid" conflicts with "accurate." In Problem 10 one must introduce a measuring device into the middle of a hardening mass which is impossible since it cannot be permitted to remain there once hard. In Problem 11 the measuring device must be introduced into a bulk material but precisely what kind of measuring device? While under the same pressure bulk materials can move with differing intensities. Problem 12 compels you to think immediately about building various mechanical devices into the wedge. One thing is abundantly clear here, namely the technical contradiction that the gain in strength necessary for extraction of the wedge is paid for at the cost of complicating the construction of a mechanized wedge.

What do these problems have in common?

They all naturally contain technical and physical contradictions. But here the resemblance ends, because the contradictions in the problems differ.

Let us compare now the inventions which provide solutions to these problems.

Solution to Problem 9: A means of finding leaks in refrigeration units filled with freon and oil (as in the majority of domestic refrigerators). In order to raise the accuracy of pinpointing leaks in the unit a luminous trace is introduced into the oil. The unit is illuminated in a darkened room by ultra-violet rays which show up the location of the leak by illuminating the luminescent element in the oil which is seeping through a hole (Pat.Cert. No 277 805).

Solution to Problem 10: "A method of determining the degree of hardening (softening) of polymer substances. In order to carry out indestructible inspection a magnetic powder is introduced into the mass and the change in the magnetic penetration of the mass in the process of hardening can be measured (Pat.Cert. Nr 239 633).

Solution to Problem 11: "An acoustic means of indicating pseudo-liquefaction of bulk materials; in order to assure direct inspection of the beginning and the intensity of movement of particles a metal sound-conducting rod is introduced into the centre of the moving material, which transmits sound vibrations which are then converted into electro magnetic signals (Pat.Cert. Nr 318 404).

Solution to Problem 12: A wedging device for consisting of a wedge and a wedge cladding designed to facilitate the removal of the wedge; the wedge cladding is made in two parts, one of which is easily melted (Pat.Cert. Nr 428 119)

Let us try to compare the specifications of the problem with what has been obtained as a result of the solution.

In the specifications of Problem 9 a substance was given (a drop of a liquid). Into the solution a second substance was added (a luminescent substance) and a field (of ultra-violet radiation) created. An analogous situation applies in Problem 10. A substance is given (a polymer), a second substance is added (a ferromagnetic powder) and a field (magnetic) set up. The self-same picture applies with two other problems: a new substance is added (a rod, cladding) and a field (acoustic, heat).

Thus it would appear that each time a substance is given one has to add a second one, together with a field. Why is this?

It is not difficult to answer this question. To enable the field to react on the first substance via the second or else the first substance by means of the second to produce at the end a field conveying information.

In point of fact it is evident that no field is capable of revealing minute drops of freon or oil. But there is ultraviolet radiation which can instead easily reveal the most miniscule quantity of luminescent particles. Thus we make up a pair consisting of a field and a second subject joining the field to the first (original) substance.

Let us designate the Field with the letter F, Substance 1 with the letter S1 the second Subject S2. We shall indicate the links by arrows. Then, for Problem 9 we can draw a diagram of the solution (the double arrow points from "what is given" to "what is obtained."

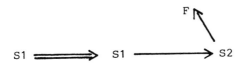

The solution to Problem 10 has the same pattern but the substance S2 itself creates the field depending on the state of S2, which in turn depends on the state of S1.

The respective diagrammes of the solutions to Problems 11 and 12 are as follows:

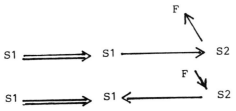

In the solutions to our problems three "active agents" are present: Substance S1 which has to be changed, processed, converted, discovered, inspected, etc; Substance S2, the "instrument" by which the necessary action is accomplished; Field F, which provides the energy, force, etc, that is, it guarantees the reaction of S2 on S1 (or their mutual interaction). It is not difficult to note that these three "active agents" are necessary and sufficient for obtaining the result required in the problem. Of themselves the field or substances can produce no effect. In order to do something with substance S1 one needs an instrument (Substance S2) and energy (Field F).

One can put it another way. In any inventive problem there is an object: in Problem 9, drops of a liquid, in Problem 10, a polymer, etc. This object cannot realise the required action on its own but has to interact with its environment or with another object. In so doing any change is accompanied by the discharge, absorbtion or conversion of energy.

The two substances and a field can be completely dissimilar, but they are necessary and sufficient for the formation of a minimal technical system which has been given the name S-Field (from Substance and Field).

In introducing the S-Field concept we utilize three terms: substance, field, mutual interaction (effect, action, connection). By the term 'substance' we understand any objects no matter what their degree of complexity. Ice and an ice-breaker, a screw and a nut, a cable and a load - all of these are 'substances'. Mutual interaction is the universal form of of connection of bodies or phenomena resulting in their mutual change. A clear description of this mutual interaction has been given by F. Engels: "Mutual interaction is the first thing we see when we look at moving matter as a whole from the viewpoint of contemporary science. We observe a number of forms of movement: mechanical , thermal, light, electrical, magnetic, chemical combination and dissolution, shifts in aggregate states, organic life, which all, if you exclude for the time being organic life, passes from one to another, mutually conditioning each other, here a cause, there an effect... (K. Marx, F. Engels. Collected Works, vol 20, p. 544).

Things are more complicated with the definition of the concept of field. In physics a field is the name for a form of matter causing a mutual interaction between the particles of a substance. One distinguishes between four different forms of field: electromagnetic, gravitational, the field of strong and weak interaction. In technology, the term "field" is used in a broader sense: there is space, to each point of which a certain vector or scalar magnitude stands in relation. Such fields are often linked with vector or scalar bearer-substances, the temperature field, (heat field), the field of centrifugal forces, for instance. We shall use the term "field" in a very broad sense, and together with the "legitimate" physical fields regard all possible kinds of "technical" fields, - heat, mechanical, acoustic, etc.as such.

In solving Problem 12 the heat field acts on S2 by altering the mechanical interaction between S2 and S1.

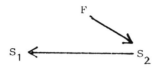

Perhaps the question will be raised, why is a heat field depicted in the S-Field formula whereas there is no mechanical field of interaction between S1 and S2? Naturally it would also be possible to depict it like this:

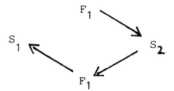

where F1 is a heat and F2 is a mechanical field.

S-Field formulae usually denote only the fields at the inlet and points, that is, fields which in the specifications of the given problem can be directly controlled - introduced, revealed, altered, measured. Mutual interaction between substances is indicated without detailing the form of mutual interaction (heat, mechanical, etc).

The accepted designations are:

$\triangle$:S-Field (overall view)

—— :action or interaction (overall view without concretization)

——→ :action

⟵⟶ :mutual interaction

----   :action (or interaction) which should be introduced according to the specifications of the problem.

⌒ :unsatisfactory action (or interaction) which according to the specifications of the problem has to be replaced.

F->   :field at the inlet: "the field acts";

⟶F   :field at the outlet: "the field responds well to the action (or change, exposure, measurement);

F'·   :the state of the field at the inlet.

F"   :the state of the same field at outlet (parameters change, but not the nature of the field);

S'   :state of the substance at the inlet.

S"   :state of the substance at the outlet.

S'-S" :"transitory" substance, finding itself either in state S' or state S" (for instance, under the effect of a transitory field);

F̃   :transitory field.

In S-Field formulae substances should be written on the line, but fields above and below it; this enables one to see at a glance the action of several fields on one and the same substance.

CONSTRUCTION AND TRANSFORMATION OF S-FIELDS

At first the notion of technical systems in the form of S-Fields encountered purely psychological difficulties. Something similar is observed in a child learning the concept of a "triangle". Why should three apples in a bag not be a triangle when those same three apples laid on a table do form a triangle? Why should three dots make a triangle, and three houses also, although the dots are very small and the houses very big? It is fairly easy to surmount these obstacles.

Incidentally, continuing the analogy with geometry, the triangle is the minimal geometrical figure. Any more complex figure, (the square, rhombus, rectangle, etc) can be expressed as the sum of triangles. It is precisely for this reason that the study of the properties of a triangle has been made into its own science of trigonometry. The S-Field - a system of three elements, S1, S2 and F - plays the same fundamental role in technology as the triangle plays in geometry. By knowing several basic rules and having tables of trigonometric functions it is easily possible to solve problems, without which it would demand laborious measurements and calculations. In exactly the same way, knowing the rules of building and transforming S-Fields, it is easily possible to solve many difficult tasks of invention.

The first rule which we have already become acquainted with, is that for greater effectiveness and controllability non-S-Field systems (a single element, substance or field) and incomplete S-Field systems (two elements, a field and a substance or two substances) have to be made up to full S-Field standard (three elements, two substances and a field).

Above we have cited Problem 3 on the separation of woodchips into wood and bark. It has two substances, and, hence, if an S-Field is to be built a field has to be added. A vast amount of research is cut down sharply. One need examine in all just a few variants. In essence, if one throws out the fields of weak and strong mutual actions (which in the given instance would clearly lead to over-complex solutions), we are left with two "legitimate" fields: electromagnetic and gravitational. Since the difference in the specific gravity of the chips is quite miniscule, one can at once reject the gravitational field. We are left with one field, the electromagnetic. Since the magnetic field does not act upon either bark or wood, we can at once set up a decisive experiment. How can chips be made to react to an electrical field? And it turns out that placed in an electrical field the particles of cork take on a negative charge, and the particles of wood, a positive. This enables one to build a separator obtaining a reliable separation of the chips.

Well, but what if the chips had not been electrified? Even in this instance the rule about the construction of an S-Field would remain in effect. The problem consists of removing one form of chip. Hence we are entitled to consider that one substance has been given which should be changed. We complete our S-Field by adding to this substance the "substance and field" pair. For example, before splitting off the trunks and the branches we apply ferromagnetic particles to the bark, and then, after the chipping process, we employ a magnetic field for the separation. Having reached this point experiments are now redundant: it is known that a magnetic field is capable of dislodging a "magnetized" bark.

This solution can be expressed as:

$$S_1 \langle\!\sim\!\sim\!\rangle S_2 \Longrightarrow S_1 \longleftrightarrow \overset{F}{\searrow} S_2 S_3$$

One is given a mix of two substances which do not lend themselves to separation. The solution consists of completing the S-Field whereby instead of S2 we take a combination (S2 S3).

The possibility of constructing "combined" S-Fields by far extends the field of application of the rule of the completion of S-Fields.

The solution of Problem 9 can also be regarded as the construction of a combined S-Field (a luminescence-bearer is introduced to a liquid):

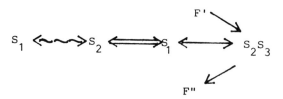

Here S1 is a refrigration unit; S2 is a coolant; S3 is a luminescent substance; F' is a field at the inlet (invisible ultraviolet radiation); F" is a field at the outlet (a visible radiation of the luminescent substance).

The Rule of Constructing an S-Field derives directly from the very definition of the S-Field concept: a barely complete technical system is known to be more effective than an incomplete system, therefore the non-S-field and incomplete S-Field systems given in the problems should be augmented into full S-Fields. Other rules exist pertaining to the construction and transformation of S-field systems. Employing these rules forms the basis of S-Field analysis, comprising one of the most important departments of the theory of solving inventive problems.

Let us take an example:

PROBLEM 13

In order to cleanse hot gases of non-magnetic dust we employ filters constituting a package formed of many layers of metallic tissue. These filters retain the dust satisfactorily but for this very reason it is difficult to clean them later. The filter has to be frequently dismantled and air blown through it in a reverse direction for a long time in order to drive the dust out. What can be done?

The problem was solved in the following way. As a filter they introduced a ferromagnetic powder located between the poles of a magnet and forming a porous structure. By switching the magnetic field on and off it is possible to control the filter in an effective manner. The pores of the filter can be small (when they are catching dust) and large (when the filter is being cleaned).

The specifications of this problem already describe an S-Field system: there is S1, there is S2 (the tissue package), there is F (the mechanical force field set up by the stream of air). The solution is that:

S2 is broken down into a ferromagnetic powder Sf;

The action of field F is directed not at S1 (the product) but at Sf (the instrument);

The field itself becomes not mechanical (Fmech) but magnetic (Fm)

This can be represented as:

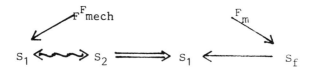

A powerful solution has been obtained thanks to the fact that one has implemented the rule of development of S-Fields: with an increase in the degree of dispersion of S2 (the instrument) the effectiveness of the S-Field is raised; the action of the field on S2 (the instrument) is more effective than the effect on S1 (the product); the electrical (electromagnetic, magnetic) fields in the S-Fields are more effective than non-electrical (mechanical, heat, etc). Really one hardly need bother proving that the smaller the particles of S2 the more flexible the control of the instrument.

It is also evident that it is more advantageous to replace the instrument (this depends on us) and not the product (which is often a natural object). In isolation the efficacy of these transformations is evident, but the strength of the rule lies in the utilization of the system of transformations.

Problem 13 has over a number of years been used as an object lesson at classes given in Public Institutes of

Inventive Creativity. In attempting it at the beginning of their course of studies, on no occasion did students give the true answer. After studying S-Field analysis, the problem was solved without difficulty by practically everyone - scientific researchers, engineers, students, schoolchildren.

Let us now return to Problem 6, which is also used widely in TSIP. Here are the notes taken by an experienced designer on the first day of studies:

"The first path is to build the requisite number of plots of land. Apparent simplicity and comprehensive results. However, in fact - expensive of realization (construction), difficulties of putting into practice. Thus this path inappropriate.

The second route is replication of only extreme conditions. The most favorable for the utilization of tractors and the most auspicious, that is, setting up on an already existing plot two lots with the corresponding soil qualities.

I take the 2nd path and, as a variant, a plot with three lots, the best conditions, the worst, and the mean."

His progress toward a solution and the answer he obtained are highly indicative of the customary designer's thinking. At first he examined the direct path - building the number of plots needed. Here there is an obvious technical contradiction: a gain in the quality of experiment is matched by a loss in terms of complexity and expense of construction. The designer goes for a compromise and makes no attempt at overcoming the contradiction. The second variant is put forward, limiting it to two or three plots. But even here there is a technical contradiction: a loss in the quality of experiment (2 plots as opposed to 200!) negates the gain in simplicity and cheapness. And once again there is no attempt at overcoming the contradiction. The second variant is presented as the most acceptable (on grounds of cheapness!) and the choice is made...

Not one of the designers tackling this problem (including some highly experienced inventors each holding some 30 to 50 patents each) could put forward a satisfactory solution. After gaining access to the TSIP (the theory for the solution of inventive problems) students at the Public Institutes (including students and schoolchildren) solved it with ease.

Here is a typical description of a solution: "Much in common with the magnetic filter problem. Sl is the soil. We introduce S2 in the form of a ferromagnetic powder. To complete the S-Field we utilize the magnetic field Fm. By activating this field we can change the characteristics of the Sl and S2 mixture.

It is interesting to compare the notes of the S-Field transformation with the descriptions of chemical reactions. In noting the chemical formulae of substances we leave aside many properties which belong to that substance. Chemical formulae say nothing, for instance, about the magnetic and optical qualities of a substance, its density, etc. They reflect only the attributes which as a matter of principle are important for chemistry: the composition and structure of molecules. In exactly the same way, in describing the S-Field formula of a technical system we leave on one side all attributes of that system apart from those which are in principle important for its development: the S-Field formula reflects the Substance-Field combination and the structure of the system.

The emergence of the language of chemical formulae became possible only when chemistry had accepted such fundamental concepts as the atom, molecule, molecular weight, and just as fundamental laws of mutual interaction and transformation of substances. Thus when equating coefficients in recording a chemical reaction we utilize the law of conservation of matter, although we do not refer to it on each occasion. Chemical unlike mathematical formulae do not permit one to discover new phenomena deriving from formulae alone and certain postulated starting points. Chemical symbols reflect only existing knowledge. In this sense S-Field analysis resembles more the language of chemistry than mathematics.

Certain inventive problems demand elimination of a harmful interaction between two objects. In these instances one must utilize the rule of breaking of S-Fields.

Let us note the formula of an S-Field taken overall.

We can break this "triangle" down in various ways: removing one of the elements, "breaking" the link, replacing the field by a third substance, etc. Analysis of a large number of problems for the breaking of the S-Field has shown that the most effective solution turns out to be introduction of a third substance, which is a variant of one of the two already available.

PROBLEM 14

In a photocopying machine a tissue paper with a drawing is spread out over the glass. To the tissue is applied a light-sensitive paper. The glass (of a special sort) breaks. Making up a new glass needs a lot of time. Therefore it is decided to put in a sheet of window glass. However, it emerges that on being adjusted the tissue becomes electrally charged and adheres to the glass. What should be done?

Engineers, not knowing the rules of the breaking of an S-Field, usually begin to select out variants which involve the elimination of electrical charges. But it is very difficult to eliminate static without blocking out the light as well and without making the equipment unduly complex. Looked at from the standpoint of S-Field analysis the problem is solved in a different way. Between the tissue and the glass one should introduce a third substance which is a variation on either the tissue or the glass. It is simpler to take the tissue, since it is cheaper. Since this tissue has to be placed between the glass and the tissue with the drawing it is necessary for it to be transparent and not retain light. This means one has to take pure tissue. Problem solved: if one stretches a clean piece of tissue over the glass it adheres. The tissue with the drawing is now applied not to the glass but to the tissue stretched over it.

This example is a clear illustration of why the rule states that the third substance introduced must be a variant of one of the two already at hand. If one simply takes some other third substance complications can set in. An "alien" substance can feel ill at ease in a technical system "outside" itself. It is necessary that there be and not be a third substance; the field is not broken, the system does not become dearer and its action is not destroyed - in other words, it does not introduce any complications. The rule of breaking an S-Field by pointing to the need for utilizing one of the substances at hand, (a variant of it), gives us a hint on how to overcome the contradiction of "there is a third substance, and there isn't"

The rule of completing an S-Field also includes a reference to overcoming contradictions. The field must act on substance S1 and the field should not (cannot) act on this substance. By introducing substance S2 and acting through it on S1 we overcome the contradiction.

In this way the S-Field analysis, like ASIP analysis, is built around the solution of problems by the discovery and elimination of contradictions.

Often one has to solve problems in which contradictions arise from the fact that one must conserve the available S-Field and at the same time introduce a new interaction. Such, for example, is Problem 8. In its specifications, an S-Field is already present, and moreover, the requisite "good" one: a mechanical field Fm

acts   via   S2   (a   disk)   on   Sl   (the   cylinder).   It   is
disadvantageous   to   remove   or   break   this   S-Field   since   the
specifications   of   the   problem   do   not   contain   any   reference
to   the   actual   process   of   polishing   itself.   Such   problems
are   solved   by   the   rule   of   the   formation   of   S-Field   chains:

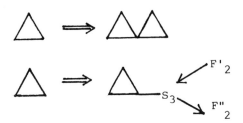

As   can   be   seen   in   the   formula,   the   essence   of   the
solution   consists   in   that   S2   (the   instrument)   is   turned
into   an   S-field   linked   to   an   existing   S-field.   Sometimes
S3   in   turn   is   turned   into   an   S-Field   continuing   the   chain.
In   problems   for   measuring   and   revealing   the   S-Field   one
must   have   at   the   outlet   a   field   which   can   be   easily
measured   and   revealed.   Therefore   in   solving   these
problems   the   final   link   Sl   -   S2-   usually   has   the   following
appearance:

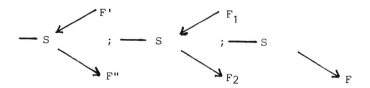

For   instance,   in   Problem   9   the   luminescence   bearer
transforms   the   parameters   of   an   optical   field   (the
invisible   ultra-violet   radiation   is   changed   into   visible

radiation):   F' => F".   No   less   often   one   encounters
transformation   of   one   field   into   another: F1 => F2.   More
rarely   one   uses   radiation   generated   by   the   substance
itself which is part of the S-Field.

If   the   substance   should turn one field into another
(or   change   the   parameters   of   a field),   one can at once
determine   the   necessary   physical   effect,   using   the   simple
rule: the   name   of   the   effect   is formed by joining the
names of two fields. For example:

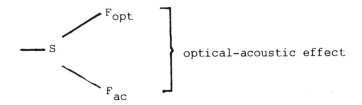

PROBLEM 15

The   shifting   of mine rock sometimes causes
a   drilling   shaft   to   "jam"   fatally   in   a
bore-hole.   In   order   to   release   the   jam,   a
vibrator   is   lowered   into   the   bore-hole to the
depth   of   the   jam.   But how   can   one   ascertain at
what depth the jam has occurred?

The   area   of   the jam is not large - just a
few   dozen   meters but the length of the shaft is
measured   in   kilometers.   The   problem cannot be
solved   by   direct   soundings; also unsuitable is
measuring   the   distortion   of   the   pipe   with a
certain   force   (the   drilling   column   cannot be
regarded   as   a   stiff   rod;   moreover the column

exerts   an   unmeasurable   amount   of   friction
against the walls of the shaft).

The   S-Field   pattern   for   solving   the problems is not
complex:

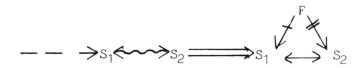

where   Fl is a mechanical field at the inlet; F2 is a field
at the outlet; Sl is the earth; S2 is the pipe.

Usually   in   solving   such problems it is expedient to
have   at   the   outlet   an   electromagnetic   field   easily
susceptible   to   elucidation   and   measurement.   As   a
transformer   substance   it   is   expedient to take the steel
pipe,   and   not   the   earth,   insofar as we do not know what
exactly   is   the   consistency   of the earth at the point of
the   jam, whereas the properties of steel are well known to
us.   Steel   is   ferromagnetic;   first   of   all   one   should
logically   make   use   of   the magnetic properties of steel.
These   properties   are already present and do not need to be
introduced   from   outside.   In   this way one has determined
the   name   of   the   necessary   physical   effect;
mechanico-magnetic   (in   physics   it   is   called   the
magnetic-elastic   effect);   the   magnetic   field   of   the
ferromagnetic   changes   depending on the current exerted by
the ferromagnet.

Inside   one   lowers   a   device   which   inserts magnetic
markers   at   each   meter. Then the pipe is winched upwards.
From   the   percussive   load all the markers above the point

of jam are demagnetized. The markers located below the jam remained unchanged. This can easily be ascertained by a magnetometer.

PROBLEMS

Try to solve several study problems. These are simple and their solution required only knowledge of the most simple rules of S-Field analysis as outlined above.

To solve a study problem: indicate the rule on which the given problem is based; give a concrete answer based on this rule. A frequent mistake consists in trying to hazard a guess at the answer using the customary method of trial and error. This is tantamount to running up to a piece of gymnastics apparatus but then walking round instead of jumping over it. It is not difficult, say, to go round a cross-beam but there is no point whatsoever in doing so. The whole sense of solving teaching problems is to work out the skills of analysis, which later are suitable for solving more difficult problems.

PROBLEM 16

The formula of an invention in Pat. Cert No 527 280 says, "A manipulator for welding work, consisting of a revolving table and a universal joint in the form of a floating mechanism, fixed by a hinged bracket to the table and placed in a liquid container; in order to increase the speed of movement of the table a ferromagnetic suspension has been introduced into the liquid and the container with liquid housed in an electromagnetic coil."

What is the point of this invention from the viewpoint of S-Field analysis?

PROBLEM 17

A pipeline with a complicated shape (bends) is used to transport tiny steel balls by pneumatic pressure. At the bends the pipe is subjected to strong wearing out on the inside by being struck by the balls being carried along colliding with the pipe walls. An attempt was made to fit protective sleeves but these too wore out rapidly.

What rule of S-Field analysis should be applied to solve this problem? What is the answer based on this rule? What would happen if, say, the pipeline were used to convey not steel but, for instance, brass balls?

PROBLEM 18

The friction of one surface against another is tested by applying a thin coat of paint to one surface and testing the evenness of the impression left on the other surface. For surfaces with higher classes of frequency it is necessary to use a very fine coat of paint (a tenth of a micron). Such a layer leaves an impression which it is difficult to distinguish.

What suggestion can you make? What rule is

your suggestion based on?

PROBLEM 19

Pat. Cert No 253 753 describes the following invention: "An electromagnetic mixing apparatus comprising a cylindrical vessel, a stator setting up an electromagnetic field and a rotor; in order to intensify the mixing the rotor is in the shape of a flexible perforated ring freely floating in the vessel." And so, instead of a rigid mixer blade, a flexible holed band is used which is brought into rotation by an electromagnetic field. Forecast a subsequent invention following up on that described in Pat. Cert. 253 753. On what rule of S-Field analysis is your conjecture based?

PROBLEM 20

In making a polishing tool one must insert small diamond grains in the shape of a small pyramid not at random but in a specific position, with the point upwards. How can this be done?

PROBLEM 21

In high-speed ships the underwater pods are rapidly destroyed because of the vibratory effect of the flow of water.

What solution can be found and on which rule of S-Field analysis is it based?

PROBLEM 22

After making certain reinforced concrete artefacts with pre-stressed (extended) reinforcement (steel rods) the need arises to

measure the stress (or the actual size of the
extension) of the reinforcement in the finished
product. The   difficulty   is   that   the
reinforcement is located inside a complete and
finished artefact. Making holes or exposing the
ends of the insert at the surface is impossible.
Scanning by means of ultrasonic waves or X-rays
is too complicated. What can be done?

3

Tactics of Invention:
Managing the Process of Task Solving

SITUATION - PROBLEM - PROBLEM MODEL

The process of inventive creativity begins with elucidation and analysis of the invention situation. The invention situation - is any technological situation clearly containing a unsatisfactory feature.

The word "technological" is used here in its broadest sense: technical, production, research, everyday use, military, etc.

Let us examine, for instance, the following situation. For making pre-stressed concrete it is necessary to stretch the reinforcement (the steel rods). In its stretched state the reinforcement is strengthened in shape and concrete is poured over it. After hardening the ends of the rods are freed, the reinforcement is shortened and the concrete shrinks, increases its firmness. In order to stretch the rods hydraulic jacks are used, but they have proved to be too complex and unreliable. An electrothermal method of stretching has been suggested: the reinforcement is heated up, a current is passed through it, it grows longer, and in this state it is strengthened. If rods of normal steel are used everything is all right, it is sufficient to heat it to 400C. in order to obtain the desired lengthening. But it is advantageous to use not normal rods but wires which can carry a greater load. In order to draw out a cable to the required length a temperature of 700C. is necessary, but on heating above 400C. even for a short time wire loses its high mechanical properties. It is unacceptable to use high-cost heat-stable wire for pre-stressed concrete .

So much for the situation. A wealth of different problems are associated with pre-fabricated concrete. In this situation we have singled out only one, stretching the wire reinforcement. It is assumed that something should be done to solve this situation. However, in the situation there is no indication as to what may be changed in the final technical system. Can one, for instance, return to the use of hydraulic jacks and try to improve them in some way? Perhaps one should improve the technology of producing heat-stable cable to bring it down in cost. Or perhaps in general one should look for a principally new method of stretching reinforcement?

The situation offers no answers to such questions. Therefore one and the same situation can generate different inventive problems..

For the inventor it is especially important to translate the situation into minimum and maximum problems.

The minimum task can be obtained from the situation according to the formula: the present situation without the drawbacks, or what there is, plus the required quality (a new property). In this way the minimal problem is obtained by applying the maximum possible limitations on alterations to the final technical system. The maximum problem, on the contrary, is obtained by the removal of constraints: one is permitted to substitute the resultant system by a principally different one. When we are given the problem of improving the sailing gear of a ship - this is a minimum problem. If, however, the problem is a phrased as: "instead of a sail find a principally different means of transport having the following characteristics..." that is a maximum problem.

It does not do to think that the transfer to a minimum problem necessarily leads to solutions of problems of lower levels. The minimum problem can be solved also on a fourth level. On the other hand, the transfer to a maximum problem does not of itself denote obtaining solutions of problems on the fifth level. Having rejected improving the electrothermal method of stretching an insert and set about improving hydraulic jacks one can certainly look to inventions of the first or second levels.

Deciding exactly whether to translate the given situation into a minimum or maximum problem is a question of strategy, and we shall return to this later. In any event, it is evident that under all circumstances it is expedient to start with the minimal problem. Its solution brings positive results, but at the same time does not demand any essential changes to the system itself, and so guarantees ease of implementation and economic effectiveness. The solution and implementation of a maximal system can take a lifetime and sometimes turns out to be entirely beyond solution given the present state of scientific knowledge. Therefore, even giving preference to a maximal problem, it is sensible at first to regard the problem as a minimal one.

Like any other problem, a problem of invention should state what is given and to what is required. A typical inventive problem looks like this:

PROBLEM 23

In preparing pre-stressed concrete a wire
reinforcement is stretched by electromagnetic
means. But on heating to the required
temperature (700C.) the reinforcement loses its
mechanical properties. How can this shortcoming
be overcome?

"Given" here includes the description of the original
technical system. "Required" includes the need to retain
everything (a minimal problem!) removing only the existing
shortcoming.

The "given" and the "required" can be presented in an
arbitrary form. The "given" can contain superfluous
information and yet omit information which is absolutely
vital. The "required" is usually formulated as
administrative or technical contradictions which are
unclear, incomplete and sometimes totally incorrect. The
solution, therefore, should begin with the construction of
a model problem simplified to the utmost, but at the same
time, accurately reflecting the heart of the problem: the
technical contradiction and the elements (parts of the
original technical system) which conflict to create the
technical contradiction.

MODEL OF PROBLEM 23

Given are a heat field and a metal wire. If
you heat the wire to 700C. it reaches the
required length but loses its strength.

First of all transposing the problem to its model removes some of the specific terminology (electrothermal method, 'reinforcement'). Stripped away are all superfluous elements in the system.

There is, for instance, no mention of the preparation of reinforced concrete: the essence of the system consists in how to stretch the wire, and the purpose for which it is being lengthened is irrelevant. Nothing need be changed, if a lengthened wire were to be used, say, for reinforcing panes of glass. Removed also is all mention that the wire is heated by electric current. The problem would be preserved were we to simply place the wire in a furnace, or heat it by infra-red radiation. All that remains are those elements which are necessary and sufficient to establish the technical contradiction.

Each technical contradiction can be presented in two ways: "If you improve A you worsen B" or "If B is improved, A gets worse". In constructing the model of the problem one must take the same formulation in which it is a question of improving (preserving, reinforcing, etc) the basic productive effect, (quality). Of the two formulations ('if you heat a wire up to 700C. it will attain the required length but will lose its strength," and "If you don't heat a wire up to 700C. it will retain its strength but not attain the necessary length" we must take the first: it guarantees the main effect (lengthening the wire) - this is precisely what the "heat field-wire" system is for.

In making the transposition from situation to problem and yet further to the model problem, the freedom of choice, (i.e. freedom to choose fruitless trials) is sharply cut down and the "outrageousness" in presenting the problem grows.

As long as we were dealing with a situation we faced a multitude of possibilities, what if we try to improve the hydraulic jacks? What if we construct a pneumatic jack? Or make a gravitational jack, in which the wire is stretched by using a heavy weight? Or if we allow for a loss in strength on heating, but then somehow restore this strength? The transposition to the modle of the problem cuts out many similar possibilities. One should preserve the electromechanical method, which has a number of advantages; one only has to remove a single shortcoming.

The following step shrinks the choice even further: we shall deliberately use a temperature of 700C., all compromises are ruled out, that is the temperature it has to be! But in spite of the natural properties of the substance we have taken, this high temperature will not ruin the wire... The problem has not only been shrunk radically, it has become "outrageous", "clearly preposterous", "unnatural". However, this all signifies only that we have rejected a huge number of trival variants and come out into the paradoxical area of powerful solutions.

In constructing the model of the problem we employ the terminology of S-Field analysis: "substance - field - action" (and with concretization precisely which of these). This allows one even before the solution to conceive of the answer in S-Field form. In fact, a heat field and substance in the model of the problem - form an incomplete S-Field. Clearly the answer will be, "it is necessary to introduce a second substance."

There are rules which permit one to build an exact model of the problem. Thus, into a pair of conflicting elements it is necessary to introduce the artefact. The second element is most often the instrument but in certain problems both elements are artefacts (for instance, in Problem 3 - the chips of wood and of bark). If one does not include the artefact in the conflicting pair, the model of the problem breaks down and we are back to square one. Take from the model of Problem 23 the artefact (wire) once again we hear the familiar refrains of the starting position: "And what if we replace the reinforcement of the concrete by something else?" Can one in general get away without extend the reinforcement?

In certain problems, it is a question of pairs of the same type of artefacts and instruments. In such situations it is sufficient to take just one pair to construct a model.

Models of the problem exclude all but conflicting elements, and not the whole technical system. Therefore the model at times can seem strange. For example, if a technical system in the problem consists of a vessel, a metal disk and a liquid acting on the disk, the model retains only two elements, - the disk (artefact) and the liquid (the instrument). Left hanging in space is the "item" of the liquid in which the disk is. In the real world this "cannot be", but the model should not be a reflection of the whole real technical system, but only a diagramme of the "weak spot" in the system.

Classifying problems (not to speak of situations) is extraordinarily difficult. The heart of the problem is concealed behind the arbitary "verbal formulations".

Models of the problem are accessible to simple and clear classification. The basis of this classification is formed by the S-Field structure of the original technical system. Such an approach permits one to instantly divide problems into three types: where one, two three (or more) elements are given. Each type is sub-divided into classes depending on which elements are given (substances, fields), how they are inter-connected and can be changed.

In Appendix No 2 is a table of the basic classes of model problems. Let us take, for example, Problem 23. In its conditions are given two elements (a heat field and a substance), i.e., the problem applies to the second type. The field and substance are connected in Problem 23 by two attendant actions: if the wire is heated it becomes longer, and loses its strength. One action is beneficial, the other harmful. This is a Class II problem.

We have not yet looked at the classification of models of problems. So far we would note only one very important circumstance. A problem of the first type (one element given) is almost always solved by completing an S-Field. Here one can draw an analogy with chemistry. Halogens possess various properties, but they have one property in common outweighing all others and conditioned by the structure of the external electron shell of the atoms of these elements: halogens strive to obtain the missing electron to complete the shell and make it complete. So it is with models of problems of the first type. Their main feature is the striving to complete a full S-Field. Problem 9 outwardly is hardly similar to Problem 6. Even from S-Field standpoints there is an identifiable difference. In Problem 9 one must find little droplets of liquid, and in Problem 6 one must change the properties of soil (and a large quantity at that!). But both problems belong to the same type of model, (one element is given) and they have continguous S-Field solutions. In order to solve both problems it is necessary to introduce a second substance and field, acting on the first substance via the second.

Problems of the third kind are transposed without special difficulties into problems of the first and second kinds. If, for instance, the specifications of the problem include an S-Field (i.e. three elements), this S-Field can be regarded as one element (a substance) and it can be joined according to the normal rules to other substances and fields.

Therefore the "classical" inventive problems are those of the second type. For a conflict one needs a clash of two contervailing tendencies, urges, properties, requirements. In its essence, such a clash is also present in problems of the first type: a second element is lacking in the conditions of the problem, but it is pre-supposed. Let us say, in Problem 20 one element is indicated, a diamond chip. A second element, which could have been indicated in the specifications of the problem is the instrument usually applied in such cases, for instance, a pincette. The chips of diamond in this instance are too small and there is no sense in even trying to grip them in a pincette, therefore the second element falls outside the bounds of the problem.

BASIC MECHANISMS FOR ELIMINATING CONTRADICTIONS

In the ASIP, four techniques are used for eliminating technical contradictions:

1) transposing the technical system stipulated in the given model problem, to an ideal system by means of formulating an ideal final result (the IFR);

2) transposing the TC (technical contradiction) to the PC (physical contradictions);

3) using S-Field transformations for removing the PC (physical contradictions); 4) applying the system of operators, drawing on the most concentrated form of information about the most effective means of overcoming the TC or the PC (lists of type methods, tables of the use of type methods, tables and indicators of the application of physical effects).

The model of the problem describes the technical system (or more accurately its 'ailing' part) and the contradiction innate to it. It is not known in advance how to eliminate this contradiction in reality, but there is always the possibility of formulating an ideal imaginary solution or final result (the IFR). The purpose behind this operation is to obtain an orientation point for the transposition toward powerful solutions. The ideal solution is by definition more powerful than all conceivable and unimaginable solutions (for the given model problem). It is as it were a solution on a non-existent sixth level. The tactics of solution of the problem with the help of the IFR (ideal final result) consist of "chaining" to this single super-powerful variant and retreating from it as little as possible.

The IFR is formulated according to the simple plan: one of the elements in the conflicting pair itself removes the harmful (unnecessary, superfluous) effect, while retaining its capacity to carry out the basic action. The ideal nature of this solution is ensured by the fact that the necessary effect is attained "for nothing", without bringing in any means whatsoever. For instance, for Problem 23 the IFR can be described as: "A heat field itself wards off destruction of the wire while ensuring nevertheless the required thermal stretching." What could be more ideal than that? Nothing was introduced, nothing has been made more complicated, but the harmful effect of the heat field has   disappeared as if by magic while the useful effect has been retained. ...The "outrageousness", the paradox, which arose already on the transposition to the model problem, is sharply increased. The heat field should not only bring about incompatible effects, but also do it itself, without any machinery, mechanisms or other devices.

In studying the theory of solving inventive problems special attention is paid to mastering the concept of the ideal machine (by which there is no machine, but the required effect is achieved), the ideal method (there is no expenditure of energy and time, but the necessary effect is obtained, and, moreover, in a self-regulating manner), the ideal substance (there is no substance, but its function is performed.)

Characteristic of ordinary engineering thinking is the willingness to "pay" for the effect required - in terms of machines, expenditure of time, energy or substances. The need for "payment" seems evident, and the engineer is concerned only that the "payment" is not too excessive and that the "reckoning" is carried out "in an educated manner." "It is necessary to fight against the transfer of heat. Well then, we will have to count on a system of heat defence. We shall use a good heat insulation, a thermal barrier, for instance. But if this is not enough, we shall have to carry the excessive heat away using heat extractors..." The inventive thought in work on the ASIP should be clearly orientated to the ideal solution: "There is a damaging factor which has to be fought against. Ideally, this factor would go away of itself. So let it eliminate itself! Incidentally, it can be eliminated, by bring it together with some other harmful factor. Or else, perhaps the most ideal thing would be to let the harmful factor begin to bring benefit..."

The tendency toward the ideal by no means signifies moving away from the reality of the solution. In many instances the ideal solution can be fully implemented. Let us say, that the ideal nature of a machine is guaranteed in that its function begins to be performed by another machine in pluralism. The ideal nature of the method is not seldom achieved by the performance of the necessary effect in advance, thanks to which at the necessary moment there is no need to waste time or energy on this effect.

The clear thrust to the ideal is necessary not only in formulating the IFR but at literally all stages of solution of the problem, in all operations using the ASIP. If, for instance, S-Field analysis suggests that one should introduce a substance, you should not lose sight of the fact that the optimum substance is when there is no substance, but whose function is performed nevertheless. There are many effective methods of introducing a substance yet without introducing it, (one substance in turn takes on two forms, the substance is introduced temporarily, etc).

The transposition to an IFR cuts off all solutions at lower levels, and cuts them off immediately and indiscriminately leaving behind the IFR and those variants which are close to it and therefore can turn out to be powerful. A further paring away of variants takes place in formulating the physical contradiction. For example, "The heat field should heat the wire in order to lengthen it, but should not heat the wire in such a way that it is spoiled."

In the physical contradiction, the "outrageousness" of the demand reaches its limit. All variants fall away apart from one or two which are maximally close to the IFR. The number of remaining variants does not exceed the number of combination methods and physical effects suitable for removing the given physical contradiction. Usually this number is no more than ten, and with the increase in difficulty of the problem the number of remaining variants decreases.

The transposition from the physical contradiction to the solution is substantially eased by S-Field analysis. Even while constructing the model problem S-Field analysis enabled us to form an general mental picture of the pathway of solution. For instance, the model of Problem 23 spoke of the field and the substance. Clearly one would have to introduce a second substance. Comparing this consideration with the formulation of the IFR one can elucidate the S-Field contradiction (SC): a second substance is needed to build an S-Field, but there should not be an S-Field if one is not to move away from the IFR. Such a contradiction (and it is often encountered in S-Field analysis) can be overcome by utilizing the "dual nature" of substances: as a second substance one can take part of the first or introduce a second substance which is a derivative of the first.

Let us take two wires, let the heat field warm one, and not the other, in doing which the lengthening of the first wire (but not the heat!) is passed on to the second wire. This is the solution to Problem 23. A heat-stable rod (which is not spent) is heated to a high temperature. The rod is lengthened. In this state it is applied to a wire. On cooling the rod shortens, and the wire which has remained cold is drawn out. For stretching one can take an ordinary wire which needs only to be twice as long as the insert, and then its temperature (for obtaining the given lengthening) can be half as much. Important is the principle of the invention - the idea of the electrothermal jack (11).

It is interesting to note that the physical contradiction is eliminated with literal precision: the heat field heats and yet does not heat the wire. Admittedly previously we had in mind one and the same wire and in the solution we are talking of two different wires. Such a "terminological conjuring trick" is performed in solving many problems. For example, in Problem 3 the talk was of separating a mixture of two identical substances. But in the solution it is proposed to mark one substance insofar as earlier these substances were in separate locations. Having been told of this solution students often say, "If only I had known that one could mark the substances beforehand.... The problem placed no ban on preparatory marking - who prevented you from knowing this beforehand?...

The simplicity of answer is sometimes taken as simplicity of the process of solution but the simpler the answer, (if it is a question of problems of the higher levels), the more difficult it can be to implement it.

Not seldom neither the construction of the model problem nor the formulation of the IFR and the PC (physical contradiction) nor S-Field analysis can provide a ready-made sufficiently evident answer. The solution of the problem needs continuation and it is necessary to go over to the operators of the transformation of the technical system. This is discussed below. So far, to sum up what has been said, we would note that following on the transfer from the inventive situation to the problem, then to the model of the problem there arises a chain of solutions: the ideal solution (the IFR is formulated) the S-Field solution (an answer in S-Field form has been found), the physical solution (the PC has been formulated and with it the physical principle of its elimination). After this should come the technical solution: working up from the idea to approximately the level of the specifications in the invention application. The process is completed by a design solution including forming the main characteristics of the new technical system. These stages - leading to a technical and a design solution - represent the progress from the solution of an inventive

problem to the designer's working out of the invention.
Here the main role is played by special knowledge and
experience. In the real creative process the "invention"
and "designer's" stages at times are oddly interwoven:
from the design stage often one has to return to the
inventing stage and amend an idea which has been found,
but in the process of construction not rarely there arises
the necessity for solving particular inventor's problems
inherent to the basic problems.

THE PROGRAMME+INFORMATION+MANAGEMENT
BY PSYCHOLOGICAL FACTORS

The process of building a model of the problem, the
statement of the IFR and the PS are clearly governed by
the second and third parts of ASIP-77 (cf. Appendix 1).
These two parts, together with the fourth which includes
the utilization of the information service of ASIP carries
the main load from the solution of problems.

Let us look at the concrete problems and how the
solution proceeds.

At Step 2.1 the specifications of the problem are
laid out without using specialized terminology. This
simple operation significantly reduces the primary
"charge" of psychological inertia. The terms are created
in order to limit the known as reliably and firmly as
possible. Any invention, however, constitutes an exit
beyond the limits of the known. If the specifications of
the problem mention, for example, increasing the speed of
an ice-breaker, then what is at first hearing an innocent
term "ice-breaker" at once commits one to a fixed set of

ideas: the need to break, smash and destroy the ice... The simple notion that it is not at all a matter of smashing ice (after all one is not concerned with harvesting ice!) and that the main thing is to move through it and not break it, this turns out to be somewhere on the other side of the psychological barrier.

At the Seed Institute Academician Lisitsyn on one occasion told the inventor Kachugin, that it was intended to hold a conference on one of the most important problems, the fight against the weevil. It was to study the conditions of life of the beetle, and in particular to determine its body temperature. At that time there was no apparatus which would have enabled them to solve such a problem.

"The budget is for fifty thousand, but we don't know whether for that amount one can construct the device you need," said the Academician.

Kachugin at once explained how to measure the temperature of a weevil using a normal medical thermometer.

Experiments had also been performed on this problem. Nineteen eighth-grade high school students were working (each one independently) on this problem for half an hour. The correct answers were given by five. Another group received the same text of the problem, but with the note: "If you replace the word 'weevil' by several simple words, the problem will become easier." As a result there were 17 correct answers in the same amount of time.

Indeed, if we substitute for the word "weevil" even the words "something very small" (a bug, a grain of sand, a drop) then the problem will be drastically simplified.

Can it really be difficult to learn the temperature of a single droplet if it is raining and one can gather a glassful of rainwater...?

Problems 24 and 25 (see Appendix 1) at step 2.1 are freed of terminology although perhaps in the specifications of Problem 25 it would do no harm to replace the term "lightning conductor" with the expression which is less attractive sounding but far more suitable for working with "conductive pole" or "conductive column."

The next step is to select the conflicting pair of elements. In Problem 24 this choice is beautifully simple: there is an artefact (a spoon), there is an instrument (a disk) - a readymade pair. Things are rather more complex with Problem 25. The specifications of the problem have mentioned a radio telescope antenna, radio waves, lightning and a lightning conductor. Sticking to the rules we try to remove the artefact and the instrument... and we come up against a rather unusual picture: in the problem there are two products (lightning and radio waves) and one instrument (the lightning conductor). Instead of one conflicting there are two unconflicting pairs: the conflict arises not "within" the pair but between them. The conducting lightning conductor does not conflict with the lightning, it is capable of "accepting" it. On the other hand, the non-conducting lightning conductor fully copes with radio waves - it does not "accept" them and so does not hold them back.

Incidentally do not be put off by these unusual expressions "conducting lightning conductor" (is this buttered butter... what other kind of lightning conductor can there be?) and "nonconducting lightning conductor" (what kind of thing is this if it doesn't conduct lightning?) One needs to have a wayout style of thinking if one is to cry out, without an ASIP: "I really need a non-conducting conductor! Not a semi-conductor, that is a conductor of a kind, but precisely a non-conducting conductor; not hot water but boiling ice, a stony gas, a dark light..." The ASIP sets the norm for this style of thinking which is not trivial or paradoxical and does not deal in contradictions. But the main thing is that this style legitimately arises as a working procedure of creative thinking: it is included not by inspiration, not by dint of chance, but according to programme, guaranteeing its steadfast maintenance throughout the whole solution of the problem.

And so, the lightning conductor has to be both a conductor and a non-conductor. According to the third rule we take for our conflicting pair a "non-conducting lightning conductor" which guarantees free passage to radio waves and the normal work of an antenna. What is this "non-conducting lightning conductor"? It can be wood, glass or a column of water. Even simpler: when one has taken away a metal column one is left with air or a vacuum, it doesn't matter.

For half a century science historians now have been telling the story of how once Paul Dirac in solving a joke puzzle about sharing out an unknown number of fish, obtained a negative number as an answer. In fact, how can a number of fishermen divide up the catch, let us say into minus two fishes (or even let the catch consist of an imaginary number of fish)... Everyone rejected such a solution but Dirac didn't, because mathematically this is a completely correct solution. The historians of science should also have asked what form of thinking it was that helped Dirac to his forecast of the existence of a positron, a "positive electron", a "positive negative charge..."

In working with ASIP negative, imaginary or else just such "unfish fish" arise automatically.

A absent lightning conductor can let radio waves pass very well, but not catch lightning. Since the lightning conductor has already been given one property (that of being absent) of the two we have compiled one conflicting pair and a technical contradiction has been obtained in canonical form. One has uncovered conflicting elements, there is a TC and the second part of the ASIP is completed by the construction of a model problem.

By moving from the technical system, described in the specifications of the problem, we cut down the number of elements examined. Now, at step 3.1 we are faced with continuing the selection process.

Of the two conflicting elements, one must be selected, i.e, the one that can be changed.

"Can be changed", "shouldn't be changed" are rather woolly definitions. Later we shall move to more accurate ones. But for the moment let simple rules suffice, quoted in the text of ASIP, which in the overwhelming majority of instances enable one to select the necessary element.

The next step is to compile the IFR. As in the previous steps, here operate clear rules, forcing one to intensify the paradoxicality of the model problem. The model demands that it should be obtained in no other way than "by itself." The ASIP does not leave one the right to think in an uncourageous way... And once more we continue the process of shrinking the field of search. Now (step 3.3) a part of the element selected in step 3.1.is separated out. Precisely to this part we are faced with "attaching" the physical contradiction, which will be formulated in steps 3.4 and 3.5.

At first glance it could seem that the steps go into too much detail in specifying the process of solution. Indeed, why should one not, for instance, join up, steps 3.4 and 3.5? Previously this is how it was done. But with time it emerged that mistakes often occur if there is an overly sharp transfer from the IFR to the FC .

If mutually contradictory demands are placed on one part of an element of the technical system, the necessity arises first of all to check whether it is not possible to "separate" these demands by simple transformations. Such a check is also carried out in step 4.1 In checking whether it is possible to separate the contradictory properties one must constantly remember the IFR: the separation should be implemented "by itself" or "almost by itself". It is not difficult to ionise a column of air; one can, for instance, use radioactive radiation. But ionised air is a conductor which like metal swallows up radio waves. It is simpler to raise and lower metal poles, and at any rate this is less dangerous for the environment. The whole point is that free charges should arise at the right moment "of themselves" and disappear "of themselves" after "catching" the lightning.

The simplest transformations provided for by step 4.1 often only map out the route for solutions in very broad brush strokes. One must do it in such a way that at the necessary moment in some way of their own volition charges occur, but in what way precisely is not evident so far.

The next step is the use of the table of typical model problems and S-Field transformations (Appendix 2).

As has been said already, the classification of model problems is founded on the following signs:

How many elements the model problem contains;

What these elements are - substances or fields;

How they are interconnected;

What limitations are imposed by the specifications of the problem on alterations to the existing elements and the introduction of new ones;

Does the problem relate to the alteration of an object (one needs to introduce a "field at the inlet") or to measurement and elucidation (one needs to field a "field at the outlet").

On the basis of these signs one can compile a detailed classification system. But many problems of the second type (two elements are given) can easily be transposed into problems of the first type, especially if there is no limitation on substituting elements. "The field interacts badly with a substance; one needs to ensure a good interaction; the field can be replaced and altered". So we get rid of the "bad" field and obtain a model of the problem of the first type (one element is given). In just the same way many problems of the third type can be easily transposed into problems of the second or third types. Therefore the table presented at the end of the book includes only those models whose transformation into simpler classes is impossible or extremely difficult.

The model of Problem 24 contains two elements (two substances): the artefact and the instrument. According to the specifications of the problem the artefact of necessity must be subjected to processing by a polishing instrument, therefore it is impossible to transpose this into a class 1 problem. The model of Problem 25 includes three elements; two fields and a substance. Once again, neither one nor the other of these elements can be removed - if the conflict disappears it destroys the model of the problem, therefore the problem belongs to class 16.

For Problem 24 the table gives in essence a readymade physical solution: the instrument should be enclosed in an F-Field, that is, an S-Field with a Ferromagnetic powder and a magnetic field dividing the substance of the disk into two substances, (one is the ferromagnetic powder), interconnected by a magnetic field. For Problem 25 the table does not yet give a final answer. By the way, here much depends on the ability to apply one's elementary knowledge of physics. Precisely apply and not just know.

Needed here is schoolboy physics, part of one's general education. Charges should appear and then disappear. Where can they disappear to? Go off somewhere? But surely they must reappear again? The physics is marvellously simple. The charges remain in situ, but are neutralized, joining up and then disuniting. The neutral molecules of air are separated into ions and electrons, and then these particles are rejoined in the neutral molecules.

Problems 24 and 25 have been "kicked around" at seminars and schools of inventive creativity for many years. In Problem 24 complications with physics never arose, after a certain amount of practice in S-Field analysis it was solved at once, "in one go". The instrument does not represent an S-Field system but according to the specifications of the problem this system can be changed and developed. That means, it is advantageous to go over to an F-Field. Difficulties usually arose with Problem 25, however. The idea of ionisation recombination is rather obvious to the physicist, but it is precisely here that a psychological barrier is erected. Ionisation in our concept is associated primarily with radiation. The idea appears of using this or another technical device generating radiation... and the solution goes into a blind alley since there is no possibility of simply and reliably determining when one ought to switch this device on.

However paradoxical it might be, the cause of difficulty is that those who solve problems (going against the ASIP) involuntarily try to make their work easier. Ionisation can be produced by the usual method, with the help of radiation (such is the voice of "common sense").

The demands of the IFR sound rather different, however. Ionisation should take place of itself. Moreover, ionisation has to be "gratuitous" and take place as if by magic at just the right time. "Common sense" shies away from such a complicating factor in the problem. The dialectic, however, is that beyond a certain point complications in the specifications of the problem actually make it easier. Let us consider once more the formulation of the IFR (now it can be expressed precisely): "On the generation of lightning when it is just on the point of "maturing" the neutral molecules should themselves separate into ions and electrons." If one takes away the word "should" we have a ready answer: we use the lightning itself as ioniser (and the thunder cloud generating it).

The IFR can be likened to a rope, clutching which a mountain climber ascends a steep incline. The rope is not being pulled upwards, but it provides support and prevents one sliding backwards. You only have to let go of the rope and you inevitably fall...

Naturally not all problems possess a solution based on elementary physics. Therefore in ASIP-77 we use the table of the application of physical effects and phenomena (Appendix 3), made up on the basis of analysis of approximately 12,000 powerful inventions with a physics bent, so to speak. Certain physical effects included in this table may turn out to be unfamiliar or little known. Then, having been given a hint by the table one can turn to the "Index of Physical Effects". Work on the "Index" was begun in 1969 at the Public Laboratory of the Methodology of Invention attached to the Central Soviet of

the   VOIR   (the   All-Union   Society   of   Inventors   and
Rationalizers).   Since   1971   the   "Index"   has   been   used   at
studies   in   Public   Academies   of   Inventive   Creativity   and   at
invention   seminars.   The   "Index"   gives   each   effect   a   short
description,   information   about   the   invention's
application,   examples   of   the   invention   based   upon   a   given
effect,   and   a   bibliography.   Especially   important   are
examples   of   inventions   -   they   allow   one   to   immediately
evaluate   the   possibilities   of   of   this   or   that   effect   and
the degree of complexity of its implementation.

In   certain   problems   the   simple   (in   the   physical
sense)   answer   turns   out   to   be   so   unusual   that   this   very
fact   prevents   its   being   noticed   and   taken   up.   In   these
instances   the   table   of   typical   methods   helps;   it   is   cited
in   (13).   In   compiling   this   table   from   a   great   mass   of
patent   information,   over   40,000   patents   and   copyrights
were   selected,   relating   to   inventions   not   lower   than   the
third   level.   Analysis   of   these   inventions   allowed
selection   of   the   more   frequently   encountered   methods   and
those   encountered   rarely,   but   always   producing   very
powerful   solutions.   These   two   types   of   method   were
included   in   the   table.   The   methods   themselves   will   be
discussed   in   detail   in   coming   chapters.   Here,   however,   we
shall   cite   only   one   example.

PROBLEM 26

On   hydratation   of   oleins   a   phosphoric   acid
catalyst   is   used   (silicon   dioxide   saturated   with
orthophosphoric   acid).   For   the   catalyst   to   work
selectively   (i.e.   specialized   so   that   it   would
produce   the   single   requisite   reaction   and   not

side reactions) it is necessary to apply heat
while making it. But experiments have shown that
on heating (even for a short time) above 250C
soluble silicophosphates appear in the catalyst
which are washed out and the catalyst loses its
activity. What can be done?

Readers who are remote from chemistry should not let
themselves be put off by the specific chemical terminology
used in this problem. It is not difficult to grasp the
essence of the problem. There is a certain substance which
accelerates the necessary reaction. Unfortunately it also
accelerates unnecessary reactions leading to the loss of
the raw material. In order that the substance speed up
only the reaction wanted it has to be heated up strongly.
But then the substance disappears altogether - it
disintegrates.

Problem 26 was examined even after the table was
compiled - to check it. The technical contradiction: the
temperature of tempering (line 17 in the table) and the
loss of the substance (column 23). Methods: 21,36,29,31.
Or the temperature: loss of time (column 25). The methods:
35,28,21,18. Method 21 is repeated: the rush principle :
carrying out the process at great speed. Heating, but
quickly and powerfully. Indeed, according to US Patent No.
3 330 313 it is proposed to "fall between" a dangerous
interval of temperatures and carrying out tempering at
temperatures between 700 and 1100. C. The catalyst loses
its activity at as little as 350., and therefore for a
long time the idea of "heating it even longer" occurred to
no one. If we heat it to 350. it loses its effectiveness,
at 500. it is quite bad... and that's all. Who would think

that from 300. another dangerous zone begins? Only one experiment was needed in all: temper the catalyst up to 1000.. But this seemed to be ridiculous and superfluous.

The table of typical methods comprises the experience of several generations of inventors, does not confine itself to "common sense." It incorporates a quality which is essential to creativity, that if "outrageous" thinking.

PROBLEMS

We include here six problems for training in the use of ASIP. It is necessary to take notes of steps 2.2 to 4.2 of the solution of this problem. Assessment of the solutions received so far is based not on the final result but only on the accuracy with which the steps are followed. If you a) have not broken the nine rules relating to steps 2.2, 3.1, 3.2 and 3.4; b) have removed the physical contradiction and c) in so doing have not brought in unwieldy constructions, mechanisms, machines and hence you have not deviated very far from the IFR, then everything is all right and the exercise is assessed as successful.

Out of habit you will come up with different solutions. (But what happens if...?) Note these answers down on a separate sheet (then they can be rejected) and return to analysis of the problem. And so, all your attention should be devoted to precisely following the steps. Calmly proceed in the direction in which the logic of the analysis takes you.

PROBLEM 27

The need often arises for measuring the

inclination of building constructions, parts of massive machines, etc. For this an inclination gauge is used, the operative part of which is a pendulum with a pointer at its end. The accuracy of such an inclination gauge depends on its length. The longer the pendulum, the greater the linear deviation of the pointer from one and the same inclination. However, an inclination gauge several meters in length is inconvenient, cumbersome (the pendulum must be mounted in a rigid housing, and dismantleable constructions are disallowed). Not acceptable also are constructions with mirrors and optical beams. The inclination gauge must remain simple but combine precision with compactness.

PROBLEM 28:

A workshop produces hollow metal cones. The size of the cones is immaterial to the problem but for the sake of precision let us take a height of 1000mm, the diameter of the lower base 700mm, the diamater of the upper section 400mm. The walls are 30mm thick. After manufacture one must check the dimensions and shape of the inner surface of the cone. It is normal practice to place one at a time a mould inside the cone (each section to be measured has its own mould). When the mould is in place it is possible to note (observing the gap) the deviation from the given shape and dimensions.

The more moulds the greater the accuracy of the measurement. But each measurement takes a lot of time and trouble. Therefore, the fewer moulds the quicker and more simple the inspection. What can be done?

PROBLEM 29

For filming and animated film a strip of drawings are made depicting the phases of movement of the object to be filmed. Each meter of film takes 52 drawings and the film is 300m long (ten minutes of screen time), representing 15,000 frames. In this way it is possible to make over 15,000 drawings and lay them out with great accuracy in such a way that the filmed movement is not shaky and does not jerk.

It is necessary to increase sharply, by hundreds of times, the efficiency of this difficult work. What can be done?

For simplicity let us consider that these are outline films (depicting only line drawings).

PROBLEM 30

The lid of a hotframe consists of a metal frame covered with glass or a stretched film. When the outside temperature goes up (let us say from 15 to 20.) one side of the lid has to be lifted in order to ventilate the hotframe. And when the temperature drops the lid has to be lowered. The angle through which the lid should rise is, say, 30..

The lid has to be raised and lowered by hand, but there are many of these hotframes and the temperature changes several times a day. The problem consists of automating the lifting and lowering process. Running an electric cable to each hotframe with a temperature sensor in this instance is too complicated and costly. The solution has to be simpler.

PROBLEM 31

Into a solid, hermetically sealed metal vessel are placed 30 to 40 cubes of various alloys and the vessel is filled with an aggressive liquid. Experiments are carried out, the aim of which being to discover how the aggressive liquid reacts upon the surface of the cubes in high temperature and sometimes high pressure conditions. Unfortunately the aggressive liquid also acts upon the walls of the vessel itself. Therefore the walls have to be made of expensive precious metal. How can this difficulty be overcome?

PROBLEM 32

A reaction glass contains a mixture of acid solutions; the work process (temperature, pressure, concentration of acids) is constantly changing. One must determine the moment at which boiling begins. Direct observation is impossible. Theoretically calculating the temperature of boiling is also impossible because of the irregular nature of the process. What can be done?

4

Talented Thought: What is it?

MODELLING WITH THE AID OF "LITTLE MEN"

With each new modification the predetermined nature of the ASIP steps increases. The provision of information is also improved. Nevertheless, ASIP does not do away with the need for thought, and only guides the process, steering one away from mistakes and forcing one to indulge in unusual ("talented") thinking operations.

Highly detailed manuals exist on flying aircraft or on performing surgical operations but learning these instructions does not make one a pilot or surgeon. Apart from knowing the manuals, practice and habits formed during practical experience are essential. Therefore, the Public Institutes of Inventive Creativity plan for around 100 hours of lectures based on ASIP and 200 hours of home assignments.

In the first hours frequent crude errors are made caused by the most elementary lack of knowledge of organized thinking. For instance, how can Problem 31 be solved? At the beginning of studies four out of five students pointed to the aggressive liquid and the walls of the vessel as the conflicting pair. The artefacts (alloy cubes) for which the technical system of "vessel-liquid-cubes" was set up did not enter the conflicting pair and hence the model of the problem. As a result a modest problem on the treatment of cubes is replaced by a far more complex one of protecting a normal metal against any aggressive liquid, including a boiling one. Such a problem, naturally, is worthy of our attention, and years spent solving it would not be wasted. The solution of such problems usually demands changes to the entire superstructure in which the system under examination is located. In these instances detailing, checking and instilling new ideas demands an enormous volume of work. Before devoting years to this (and possibly even one's whole life) it is worth while spending five minutes solving a simpler but also necessary problem: how can one do something about the cubes themselves?

If we were to take the "cube-liquid" as the conflicting pair the chamber would not enter into the model problem. At first glance this makes the conditions more onerous. Since it is not a matter of the walls of the chamber they can be made of anything (or may not even be there at all!); one has to search for a solution in which retaining an aggressive liquid is not dependent on the walls of the vessel at all... As usual making the problem temporarily more difficult in fact means making it simpler. In point of fact, what is the conflict now that one is left with the "cube-liquid" pair and the "chamber" is left "out of court"? Is it the aggressive action of the liquid? But surely in this combination the liquid is supposed to be aggressive, that is its useful (its only useful!) characteristic... The conflict now is that the liquid will not (without a chamber) stick to the cube but will simply disperse, leak and flow away. What can one do to prevent the liquid flowing away but keep it reliably in the vicinity of the cube? Pouring it into the cube is the only and rather obvious answer. A gravitational field is exerted on the liquid, but this effect is not transmitted to the cube and therefore the liquid and the cube do not interact (mechanically). This is the simplest problem on building an S-Field: let the gravitational field act on the liquid and transmit this effect to the cube. Replacing the cubes by "glasses" (hollow cubes) is the first thing that comes to mind if the cube and the liquid are taken for the model problem instead of the liquid and the chamber. There is a wall (the wall of the cube) and there is no wall (the wall of the chamber) - an exemplary elimination of a physical contradiction. Such a solution quite clearly has no need of being checked - it is absolutely clear and reliable - and one needs no draughtsman's drawing board and there is no problem of implementation. And in order to arrive at this solution no more is necessary than meeting a direct and simple prescription of ASIP: the conflicting pair should contain an artefact and directly working on it an element of the system. Or (as in the lightning conductor problem) one could examine the conflict between two pairs, "cube-liquid" and "liquid-chamber". The IFR is missing, since the liquid does not act on the chamber itself, preserving its capability of acting on the sample. Here the path toward solution is once more shorter, for from the very beginning it was accepted that the liquid is missing. At once a clear contradiction occurs: there is a liquid (for the cube) but there is no liquid (for the chamber). According to the specifications of the problem, one must not separate the conflicting properties in time (the liquid must act without interruption upon the sample); there remains one possibility: to separate the conflicting properties in space - the liquid is where the cube is but not where the chamber is.

The text of ASIP-77 includes nine simple rules but alas, it is not so simple to teach their implementation. At first these rules are ignored, "they pass straight through", then they begin to apply them incorrectly and only gradually, somewhere in the second hundred of problems is the skill worked out to work confidently with the ASIP. All teaching is difficult but teaching the organization of thought in solving creative problems is doubly difficult. If one sets the problem of calculating the volume of a cone a man can write the formula incorrectly, multiply it incorrectly, but he can never say without even looking at the figures: "The volume of a cone? But what if it were equal to 5cm3 or 3m3 What colour is the cone painted? And what if it is not a cone at all? Perhaps it would be better for us to calculate the weight of some kind of hemisphere..." In solving inventive problems such "pirouettes" are essential to "quest for a solution" and disturb no one in the slightest...

There are many fine techniques of solution which even today cannot be put into simple rules. They have not yet been included in the text of the ASIP, but they can be "built-in" on the judgement of the lecturer when the students are accustomed to performing analaysis without breaking him off somewhere in the middle with the age-old "But what if we do it like this...?"

As we have already said, in creating synectics Gordon added to the brain storm three types of analogy, including empathy - personal analogy. The essence of this method is that in solving a problem a man "enters" the

form of the object to be improved and tries to realise the
action required by the problem. If in so doing he succeeds
in finding some kind of approach, some new idea, the
solution is then "translated" into technical language.
"The essence of empathy," says George Dixon," consists in
"becoming" an object and from its position and its
viewpoint, looking to see what can be done" (9, p.45).
George Dixon goes on to point out that this method is very
useful for obtaining new ideas.

Practical experience in using empathy in solving
teaching and production problems shows that empathy can
sometimes really be useful but sometimes harmful too. Why
is this?

In identifying himself with a particular machine (or
a part of it) and examining possible alterations to it,
the inventor involuntarily selects those which are
acceptable to man and rejects any which are unacceptable
to the human organism, such as dissecting, splintering,
dissolving in acid, etc.

The indivisibility of the human organism prevents one
from successfully employing empathy in solving many
problems as in, for instance, Problems 23-25. The
drawbacks of empathy are overcome in modelling with the
help of little men (MLM) - a method utilized in ASIP. In
essence it amounts to presenting the object in the form of
a multitude (a "crowd") of little men. Such a model
preserves the merits of empathy (making things visibly
obvious, simplicity) and does not have the drawbacks
implicit in it.

Instances have been known in the history of science
where something akin to MLM has been spontaneously

applied. Two such instances are especially interesting. The first is Kekulle's discovery of the structural formula of benzole.

"One evening I was in London," Kekulle relates, "and was sitting in an omnibus and thinking about how to depict the molecule of benzole C6H6 in the form of a structural formula answering the properties of benzole. At that time the bus passed a cageful of monkeys which were chasing and catching hold of each other, breaking the chain and once more linking up to form a circle. Each one held to the chain with its back paw, and the next one held on to its other back paw with both forepaws, waving its tail merrily about in the air. In this way five monkeys by holding each other formed a circle and the thought immediately crossed my mind that this was a representation of benzole. Thus arose the formula quoted above, and it explains for us the firmness of the benzole ring. (op cit 7 vol 2, pp 80-81).

The second instance is even more well known. This was a mental experiment of Maxwell while working out a dynamic theory of gases. In this mental experiment there were two vessels with gases at the same temperature. Maxwell was interested in how to obtain rapid molecules in one vessel and slow molecules in the other. Since the temperature of the gases was the same the molecules would not divide of themselves: in each vessel at any point in time there would be a specific number of both rapid and slow molecules. Maxwell mentally joined the vessels by a tube fitted with a valve which was opened and closed by "demons", imaginary beings approximately of molecular dimensions. The demons let rapid particles from one vessel into another but closed the valve to small particles.

These two instances are interesting primarily since they explain why the MLM used little men and not, for instance, little balls or microbes. For modelling it is important that the small particles should see, understand and be capable of performing actions. These demands are most naturally associated with man. He has eyes, a brain and hands. By using MLM the inventor uses empathy at microlevel. The powerful aspect of empathy is preserved, without the shortcomings inherent in it.

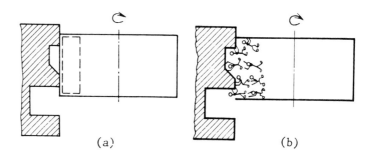

(a)   (b)

The Kekulle and Maxwell stories have been retold by many writers but none of them has linked them together or pondered over applying these two instances to different branches of science and converting them into a method to be deliberately employed. The Kekulle story is usually quoted in order to illustrate the role played by chance in science and discovery. And the Maxwell experiment is quoted in order to draw the conclusion which one knew anyway that the scientist needs an imagination...

The technique of utilising the MLM can be reduced to the following operations:

-at step 3.3 one needs to isolate part of the object

which cannot fulfill the demands indicated at step 3.2 and imagine this part in the form of little men.

-One needs to divide the little men into groups behaving (intermingling) according to the specifications of the problem:

-the model obtained should be examined and reconstructed in such a way that conflicting actions are performed.

For instance, in Problem 24 the diagramme for step 3.3 usually appears as depicted in fig. 1,a: one sees the outer layer of the circle which structurally differs in no way from the central part of the circle. In fig. 1,b the same drawing is shown but using MLM. The little men coming into contact with the surface under treatment remove particles of metal, and other little men keep back the "workers", preventing them from flying out of the circle, falling or being thrown off. If the depth of the cavity is changed, the little men are reorganized accordingly. Looking at the left hand drawing and not simply jumping to the conclusion about the necessity for breaking the outer part into "grains", having made these grains movable and at the same time "chained" in a circle. The right hand drawing leads to this idea.

Once at a TSIP seminar the students were set the problem of increasing the speed of an icebreaker. It was impossible to achieve this by increasing the power of the engines; modern icebreakers are so "crammed" with engines that there is almost no room left for useful cargo (similar conditions to the problem and description of the ASIP solution cf (13,pp179-188).

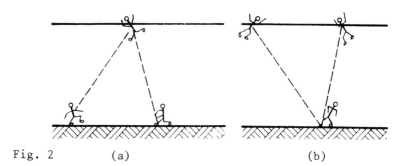

Fig. 2   (a)   (b)

At first they tried to solve the problem by using empathy. One student putting himself into the "image of the icebreaker" paced the room deep in concentration, and went up to the wall. "This is ice," the student said. "and I am an icebreaker. I want to pass through the ice, but the ice does not let me through..." He banged up against the "ice" taking a running jump at it, at times the feet of the "icebreaker" tried to pass under the wall, but his torso prevented this, sometimes the torso tried to pass over the wall, but his legs prevented him. By identifying himself with the ice breaker the student transferred to the icebreaker the indivisibility characteristic of the human organism, and by so doing complicated the problem, empathy in the given instance only hindering the solution.

At the next study session the same student solved the problem by using the MLM. He approached the table, thought for a few seconds and then said rather distractedly, "I don't understand, what the problem is... If I consist of a crowd of little men, the upper half of the crowd will pass above the table, and the lower part beneath it...Evidently the problem now is to join both parts of the ice breaker, the above water and the under water part. We shall have to introduce some kind of bars, which are narrow and pointed, these would easily pass through the ice and we would not have to break up a huge mass of ice..."

The MLM method has not yet been exhaustively researched, and there is a lot that is puzzling still in it. Let us say in problems to measure the length of a separated part of an element, it is better to present it not in the form of an unbroken of men, but as "every other" rank. Even better, if the men are placed in the form of a triangle. And better yet in an irregular triangle (with unequal or bent sides). Why? So far one has only been able to form guesses. But the rule works...

Let us take just Problem 7. One must measure the depth of a river from an aircraft. According to the conditions of the problem a helicopter cannot be employed, landing men is not allowed and it is also impossible to utilize some kind of properties of radio waves, because there is no chance of ordering special equipment. Morever the depth survey should be carried out at virtually no cost (admissible is only the expense of paying for a flight along the river.)

Let us use the MLM method. As yet unknown is the "measuring rod" we are to employ, having thrown it or directed it from the aircraft in the form of an irregular triangle. Conceivable are only two variants of the location of the little men (figure 2) forming this "measuring rod."

The upper little men should be lighter than water, the lower heavier. Let us suppose that these are pieces of wood and stones joined by a line (fig. 3).

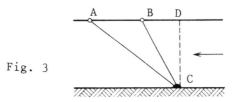

Fig. 3

It is not difficult to create such a triangle. Pieces of wood A and B are joined to stone C by lines and it is known that the length of both lines is greater than the depth of the river (this can be checked by a simple cast). The deeper the river the greater the distance AB (the pieces of wood not joined together). Moreover, to one of the floats should be fastened (as a "scale") a metre rule and it is possible to drop this "device" and then photograph it from above.

Knowing AC and BC and having measured AB from a photograph it is easy to calculate CD. The solution is remarkably simple and elegant (Pat. Cert. No. 180 815. Arriving at it without being given a hint ("Drop three little men, order them to take up positions in the shape of an irregular triangle...") is very difficult, the reader may be convinced of this by putting the problem to his colleagues.

Let us turn now to Problem 8 in which it is a matter of measuring the radius of a polishing disk and where one can also cal upon little men.

The polishing disk is applied to the part and the process of polishing is all right (unlike Problem 24), since the S-Field already exists. But the disk is operating within a cylinder, and one must determine the length of the radius of the disk without withdrawing it from the part. This is a class 14 problem. Its solution (according to the table of typical models) is: to S2 one must join S3 which is changing field F and depending on the state of S3 and consequently of S2. If we attach an electrically conducting strip to the hub of the disk and pass a current through it, then by measuring the changes in resistance we can assess the changes in the radius of the disk (fig. 4).

Unfortunately such a plan does not ensure accurate measurement. Resistance depends not only on the length of the strip but also on the force of pressure of the disk on the surface being polished, on the state of the "chain-shaft" contact and on the temperature of the disk...

Let us try to dispose the little men in an "alternating" chain (fig. 5).

Now the dimensions of the radius of the disk can be assessed by the number of impulses of current and the size of the impulses themselves has no significance. The solution is far more effective than the previous one. Admittedly putting a current through each little man is not so simple.

Let us turn to the "triangle". A regular "triangle" offers us nothing. Instead an irregular one provides one more solution (fig. 6), but now with defects no longer: with the change in the radius changes the permeability (the relation of signal to pause) of the impulses passing through which permits one to measure simply and reliably the radius of the disk. The MLM method has other not so obvious wrinkles. The time will come when we will take the laws in operation here and the method will be made a part of the ASIP as mandatory steps. This is what happened, for instance, to the DTC operator which at first also seemed strange and exotic.

DTC stands for Dimensions - Time - Cost. Any technical system given in the specifications of a problem takes a form which is familiar to us.

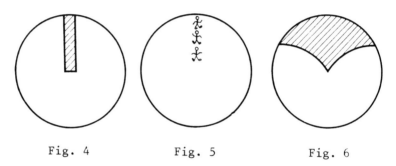

Fig. 4   Fig. 5   Fig. 6

One can, for instance, take the word "icebreaker" out of the context of the problem, but it still retains the image of an icebreaker, something "in the shape of a ship", approximately corresponding to the size of an icebreaker, acting approximately at the same speed and costing approximately the same. The term does not exist yet, but the form of the original system has been preserved and carries a strong charge of psychological momentum. The aim of the DTC operator is to overcome this inertia, to smash the obtrusive old form of the technical system. The DTC operator includes six mental experiments, reconstructing the conditions of the problem (step 1.9 in the ASIP-77 text). The experiments can be implemented on different levels, here much depends on the strength of imagination, on the nature of the problem and on other circumstances. However, even the formal fulfillment of these operations cuts deep into the psychological inertia connected with the usual form of the system.

STRUCTURE OF TALENTED THOUGHT

A powerful imagination permits a more effective use of the DTC operator. But the very fact of using it in turn develops the powers of imagination. I have already stressed frequently that the ASIP does not simply organize thought but also organizes talented thought.

But what then is this talented thought?

We shall apply ourselves to this problem again.

PROBLEM 33

There is a speed boat which holds the absolute speed record. It possesses the ideal shape, and the very best engines. How can one set a new record which by far exceeds (by 100 to 200 km) the existing figure?

The imagination of the run of the mill inventor will obediently sketch in the record holding speed boat. He switches on a mental screen on which a clear picture appears. Onto this initial picture the imagination begins to make various changes. For a long time the weak inventor examines variants and things move very slowly along. The variants (even the tenth or fifteenth) differ only slightly from the original. "Perhaps one could lengthen the hull? Should one give the hull a more streamlined shape? Can one build in a more powerful engine?"

A powerful inventor selects the variants more boldly. On his mental screen the pictures rapidly alternate, and unusual pictures are featured. Variant 67: "What if we cover the boat in something like a leopard skin: after all it isn't by chance that the leopard runs faster than all other land animals. Perhaps his fur helps him retain a streamlined smoothness, and prevents eddies from forming? (Incidentally, recently the Soviet inventor G.N. Sutyagin was awarded Pat. Cert. No 464 716 for a "surface made streamlined by means of a liquid or gas".

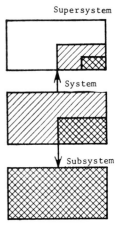

Fig. 7

The description of the invention says: "... with the aim of lowering resisitance put up by friction... its face (surface) is filled out with artificial fur, fleecy cloth and similar materials").

Technical systems exist not in isolation. Each one is part of a supersystem of which it forms a part reacting with other parts. But the systems themselves also consist of interreacting parts and subsystems. The first indicator of talented thinking is the ability to shift from the system to the sub- and supersystems and for this three mental screens have to be at work (fig.7).

In other words, when looking at a tree (the system)

we have to see the wood (the supersystem) and individual parts of the tree (the roots, the trunk, twigs, leaves, etc) Incidentally, not only at this, but at each stage it is necessary to see the line of development: the past, present and future (fig.8).

What does this mean, "see the line of development"? Take one of the subsystems of the speed boat, the hull. The greater the speed the greater the resistance offerred by the surrounding medium. And for this reason the hull tries to shrink, get smaller. The ideal hull would be when there is no hull at all... And the engine, the other subsystem of the boat on the other hand tries to get bigger and more powerful. If one were to give it its way it would fill the whole hull and then some, breaking to the outside. The struggle of these two mutually contradicting tendencies also determines the line of development of the boat's susbsystems: the hull contracts, shrinks and becomes even more "willowy", whereas the engines swell and grow, filling all the empty space inside the hull.

Great passions are constantly clashing on the mental screens of the talented thinker. Contradictory tendencies collide, conflicts arise and peak, opposites fight it out... In the heat of this struggle the imagination sometimes changes into an anti-imagination.

Together with the speedboat appears an anti-speedboat. The usual speedboat floats, hence the anti-speedboat doesn't. A ship which cannot stay on top of the water sinks... Seen from the viewpoint of normal thought processes this is pure stupidity.

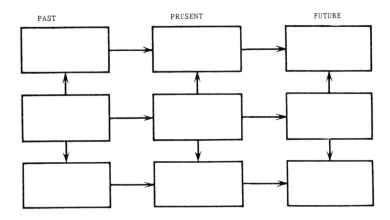

Fig. 8

But what if we give it some thought nevertheless? The
"mean density" of a normal ship is less than a one, and
this is precisely why the ship stays up. Inside the hull
is a lot of free space, hence the large volume of the ship
and the great resistance it puts up to the surrounding
medium when it moves. Underwater pods admittedly lift the
hull above water, but there is still air resistance.

The anti-ship is not obliged to stay above water.
Consequently it can be filled to bursting point with
"iron" - the engines. The greater the power of the
engines, the greater the speed. But with its beautiful
super-powerful engines the anti-ship would sink like a
stone to the bottom... Incidentally on moving it would kept
up by virtue of the lifting force exerted underwater
wings. And at its moorings it could use as "floats"
auxiliary inflatable containers. At its moorings our
anti-ship like a normal ship (or dirigible) would stay
afloat by Archimedes Principle. But once it had got up

speed, and having lifted its hull above water, the anti-ship would "contract" by jettisoning its now superfluous auxiliary containers (the dirigible would become an aircraft).

The idea of an anti-ship no longer seems so far-fetched. On the contrary, it is the normal structure that seems strange, in which the hull kept above water retains a large volume needed only in the water...

In 1911 the Wilson chamber was built - one of the main instruments of nuclear physics. The charged particles moving in supersaturated water vapour filling the chamber became visible and formed a trace of droplets of liquid. Thousands of improvements to the Wilson chamber have been put forward, but for almost half a century no one thought of the idea of an "anti-chamber" in which a trace would be formed by bubbles of gas in a liquid environment. In 1960 D Glezer received the Nobel Prize for creating a bubble chamber...

But to return to the screens of talented thought. Three stages, nine screens, depiction and anti-depiction - this is after all an elegant simplification of the pattern. A genuinely talented thought has many stages above the system, (the supersystem, the super-supersystem-...) Behind the tree one should see not only the wood but also the biosphere in general and not only the leaf but also the cell of a leaf. Many screens should be to the left of the system (the recent past, the distant past) and to right of it (the near future, the far-distant future). The depiction on the screens becomes now large, now small, now the action slows down, now it speeds up...

Complicated? Yes, indeed. The world in which we live is constructed in a complicated way. And if we want to learn about it and transform it, our thinking must reflect this world correctly. The complex, dynamic, dialectically developing world should find in our consciousness its full model which is complex, dynamic and dialectically developing.

A mirror reflecting an image of the world should be large and many-sided. Just as in Ciurlione's paintings.

Perhaps no other artist possesses such a strong "systematic vision" of the world. Many of Ciurlione's paintings give on one canvas not only the "depicted system," but also its "subsystem" and its "supersystem" into which the "system" fits. The Sonata of the Sea (Allegro) shows three different scales at the same time. From a bird's eye view are depicted hills on a coastline. But the waves are drawn to another scale. They are shown with the eyes of a man, standing in the shallow water; through the water are seen the play of light and shadow on the sandy bottom and the silhouette of fishes. And here there is yet another scale, quite massive, for the "subsystem": drops and bubbles of air are seen almost in close up.

The reader is right to ask: surely here you are talking not about talented thought but about genius. Precisely so. What is more even among geniuses such thinking is by no means encountered every day.

In reality "the full screen pattern" shows the thinking of a genius in his starry hours, which are extremely rare even in the lives of great thinkers and artists. "The full picture" is the IFR and approaching this idea is the ASIP. It is not difficult to note that the ASIP represents a linear evolution of the "full pattern" plus the informational guarantee, allowing us to "draw" the "depiction" required by the pattern.

THE DIALECTICS OF ANALYSIS

In studying TSIP individual operations are mastered at first comprising the "full pattern", and then the most difficult part begins, joining up the separate operations into one system of thought. At this stage together with the solution of usual inventive problems training complex problems are needed. In particular in experiments the question is used: "What is the sense in life?"

If a group has only just started its studies, the usual selection of variants takes place: all are variants on the level of the original system ("the sense of man's life") and only at the present time.

fig 9 rp 70

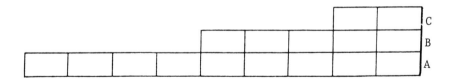

Or else the session is held with a trained group. At once correctives are introduced to the very presentation of the question. Life should be examined at a minimum of three

levels:(the cell, the organism, the society), and in so
doing at each level there should be three stages (the past,
present and future). A pattern emerges which is similar to
that cited in fig. 8. But the cells are more ancient than
organisms and organisms more ancient than society, the
pattern needs to be changed, that is clear (fig. 9)

Fig. 10

The development of single cells slowed down since
nature "invented" the organism (stage B). Second
correction: the (biological) development of organisms has
slowed down since the time that society was "invented"
(stage B). The main line of development proceeds in steps,
passing from stage to stage (fig. 10).

The pattern can be complemented from below by even
longer stages: the "lives" of molecules, atoms, elementary
particles... The heaviest atoms are unstable: the "stage"
of atoms is interrupted somewhere around the hundredth
"model", and further development takes place at the expense
of combining atoms into molecules. The "stage" of molecules
intercepts the relay-race of development. More and more
complex molecules are formed, reaching up to polymers and
proteins. However, with the appearance of proteins the

development of molecules stops: the relay race has been intercepted by cells, which also form a "stage", consistently following their "models", and although very large cells are known (among the water-plants) the development is once again intercepted by the supersystem - the organism. A very simple uniting of cells takes place, but gradually ever more complicated organisms arise, right down to man. Incidentally, long before the appearance of man nature had begun to "experiment", trying to make out of organisms (bees, ants) supersystems. Evidently these experimental supersystems turned out to be bad for one decisive reason: they did not guarantee an acceleration in the rate of development and on the contrary the rate of development of these supersystems turned out to be as good as nil. Nature was forced to "invent" man, and only then did development enter the next stage.

The question arises as to the reasons for the "ladder relay-race." The answer is almost obvious. The higher the stage, the more it is independent of outside conditions. Elementary particles (if they interact with the outside environment) have a miniscule life. Neo-organic (and simple organic) combinations are more "living" but they too are almost defenceless against outside influences, heat, cold, chemical reactions. The protein and the cell are higher degrees of organic matter in its struggle for independence from outside conditions. An even higher stage is the organism. The cells of our body are renewed on average every seven years; the organism as a whole lives usually longer. It is restored even in those cases where outside influences destroy part of the cells. Society is even more persistent in regard to outward effects and far more protected than the individual organism.

It is curious to apply this construct to the analysis of a Solyaris by Lem or Hoyle's Black Cloud. In both cases there is a clear violation of the "ladder" relay; the organism should pass to the level of a society, but it continued to increase, remaining a single organism, and grew to assume the dimensions of an entire planet.

This pattern can also be extrapolated upwards. The development of society can go for a certain time, and then comes a transposition to the next "stage", in which society will play the same role as the cell plays in the organism.

Now much play is being made of extraterrestrial civilizations. What are they like, these civilizations alien to earth? Why do they not seek us out and make signals to us? Why do we not see any manifestations of their activity?

A supercivilization is thought of on the level of a society, but only as a more developed, more energetically equipped. But in fact a supercivilization has to be a stage higher, on the super-society level. Can the individual cell take account of what the organism will specially seek it for (for establishing contact!).

More and more resources and efforts are being pumped into radio telescope projects and attempts to pick up signals from super civilizations. But meanwhile from the pattern it is seen that each stage more and more rapidly creates the conditions for the appearance of the next stage. Above the "society" stage should comparatively quickly appear the "super-society" stage and then, even more quickly, the "super-super-society". The super-civilization can turn out to be more remote from us (in terms of stages) than man is removed from elementary particles.

Note that we have not yet begun our investigation of the problem ("What is the purpose of man's living?") but the very process of placing the question in the "full pattern" has already given us much that is new and interesting. One should stress, this is all but a fragment of a single study. The academic program of the public schools of inventive creativity includes 15 studies of this type, together making up a course in the development of the imagination. Other examples the reader will find in (19, pp 138-166).

As a result of such studies the mechanism has become clear for the development of technical systems, in particular the "stepped" nature of this development. Having exhausted the reserves of development, the technical system becomes a subsystem as part of a more complex system. When this happens, the development of the primary system sharply slows down. The baton is picked up by the system being formed.

Let us take just the story of shipbuilding. Ships brought into motion by oars gave way to sail and oar-propelled ships and oars ceased to develop. There began the long life of the new system - sail and oar. Gradually they became sail only, and then again began a changeover to a more complex system, sail and steam. The rate of development of sails slowed down, and with time the sail-steam vessels became steam only...

DUNKER'S EXPERIMENT

So we know of many signs of talented thinking and we have a firm foundation on which to judge whether the operations carried out in solving problems are good or bad and whether they will lead us up a blind alley or nearer to the answer. But is it possible that psychologists experimenting on the solution of problems didn't know about the systematic approach, about the IFR, etc? How on earth did they conduct their experiments?

Let us take one of the most famous experiments of K. Dunker - his solution of a problem of X-rays (1926). In 1935 another work of his appeared, more detailed but based on the same experiments.

This was his problem [5, p. 49]. "Your problem is to determine in what way one can use a certain form of X-rays having a greater intensity and capable of destroying healthy tissues, in order to cure a man of a tumour in his organism (for instance in his stomach)".

Below we cite the record of the solution which, as K.Dunker writes, is "especially rich in typical thought processes" (5,p.88):

1. Introduce the rays through the food passage.

2. Render the healthy tissues insensitive to rays by introducing chemical substances.

3. By operative means remove the stomach to outside the body.

4. Reduce the intensity of rays when they pass through healthy tissue, for instance (can this be done?) switch the rays on fully only when they reach the tumour (the Supervisor: a false concept, rays are not a hypodermic syringe).

5.Take something inorganic (which does not let rays

pass through) and thus protect the healthy stomach walls. (Supervisor: You have to protect more than the stomach walls).

6. Something like: either rays have to penetrate to within, or the stomach has to be outside. Perhaps one should relocate the stomach? But how? By pressure? No.

7. Introduce a tube (into the abdominal cavity)? (Supervisor: This is generally done when one must use some agent at a definite place to produce a certain reaction which must be avoided on the way there?)

8. Neutralize the effect on the way there. I am trying to do this all the time. 9. Remove the stomach to outside the body. (Supervisor repeats the problem and underlines: "given a sufficient intensity.")

10. The intensity should be such that it can be changed.

11. Harden the healthy tissue by preliminary weak irradiation (Supervisor: How can one ensure that the rays destroy only the vicinity of the tumour?

12. I see only two possibilities: either protect the healthy tissues, or render the rays harmless. (Supervisor: How can one minimise the intensity of the rays on the way to the stomach?)

13. Somehow deflect their diffuse radiation - dissipate it. Pass a broad and weak beam of light through a lens in such a way that the tumour is in focus, and consequently, under the powerful effect of the rays (The general duration around 30 minutes)".

And so, more than ten trials have been carried out. In 30 minutes we have drawn near to an answer (In the tumour many weak rays coming from various directions intersect.)

In this process the Supervisor has on many occasions intervened during the process of solution.

Let us introduce one rule of the ASIP: one may have to change the instrument, but not the artefact (which is a technical and not a natural object). Let us examine each stage of the solution taking this rule into account.

1. In the problem there are two substances (the tumour and the healthy tissue around it) and one field (the X-rays). Both substances are natural, both are artefacts. The rays are the instrument. The first variant is an attempt to do something with the healthy tissues (find a "passage through" them. This is a clear violation of the rule and hence it is a futile variant.

2. Again an "artefact" is taken as the object, again it is an empty variant.

3. An "artefact" is selected - an empty variant.

4. For the first time an instrument is selected! The formulation, by the way is near to the IFR. But the supervisor crudely breaks off an excellent idea. The examinee has formulated very well what one needs in the ideal case. The rays, like an injection needle, at first cover a large volume (irradiating the whole tumour). Much - little - much. The supervisor should have said that at last the necessary element (the rays) have been selected, but now we only have to think about them. Incidentally, the supervisor hinders the examinee and brings him from the right path by saying that rays are not a syringe and that the idea is unsuitable. Can one arrange it so that the density of energy is different along the length of the ray? In principle one can: stationery waves; a loop in the area

of the tumour. From here incidentally it is easy to arrive at the idea of a pencil of rays.

5. The examinee, driven by the supervisor from the right path once more turns to an element, which it is impossible to change.

6. Translated into the language of ASIP, what do you go for, the instrument or the artefact? Having put the question this way, the examinee goes for the artefact.

7. Again an artefact is chosen. And the supervisor begins to give the examinee a push in the right direction, drawing his attention to instrument. designated by the word "agent".

8. The examinee argues quite reasonably that one must neutralise it (here there are two ways - making the tissue insensitive or somehow rendering the rays harmless.)

9. Again an artefact is chosen. The supervisor is forced to draw the attention of the examinee to the rays. For this he has to repeat the problem and stress the words relating to the intensity of the rays.

10. The examinee partially returns to his fourth formulation.

11. But here once more he leaps to that element which cannot be changed. Then the supervisor, abandoning all subtlety, directly "turns" the examinee's "face to the rays".

12. The examinee repeats the old question "either - or". And here something staggering happens. The supervisor makes a straight leap, repeating what he himself has rejected in Step 4: how to reduce the intensity of the rays on the way to the stomach?

13. Naturally the correct answer is obtained at once.

By knowing the rule (change the instrument) we now see what is the reason for the mistakes of the examinee and supervisor. Perhaps on this occasion the supervisor reacted worse than the examinee. The examinee had formulated something approaching the IFS and the supervisor brusquely knocked him away. The supervisor wanted a direct approach to the answer and he did not take account of the fact that the path of knowledge is not a straight line. At first a digression is needed, to the IFS and from there to the answer.

What does K. Dunker himself do next? How does he analyse the record? Like this. He arranges the answers in three groups:

1. Elimination of contact between the rays and healthy tissues (in the language of ASIP he examines two elements, the artefact and the instrument).

2. Lowering the sensitivity of healthy tissues (he examines the artefact).

3. Lowering the intensity of the rays on their path through the healthy tissues. Two variants relate to this, 4 and 13 (an instrument has been taken as object. An interesting point: the logic of analysis has forced Dunker to join the answers 4 and 13, but he has not reviewed his own answers to variant 4 and has not seen that it led the examinee away from the right path.)

Thus 11 of the "empty" variants would not have appeared if the rule about the artefact and the instrument had been introduced.

This problem was placed before a 15-strong group of students (at a Professional-Technical Academy). One student solved it (a cinema technician for whom the dispersal and focussing of rays are a piece of cake), and for 45 minutes the others did not come up with the model answer. After explanation of the rule, it was solved by everyone and the longest selection went to four variants. In another group the rule was explained first and then the problem was given. All solved it and over half with their first variant.

Why then did Dunker not notice what emerged so distinctly once he had grouped the variants? Why did he not discover that the mistakes were associated with attempts to change natural objects and correct answers were linked to changing the instrument? Dunker is a psychologist. He is interested not in the objective laws of the development of technical systems, but in the psychological aspects: how an examinee elucidates a problem, how the solution evolves (from the first idea to the final formulation) etc. Dunker, like other researchers studying creativity from "purely psychological" positions, did not understand that the development of systems is primary and psychology secondary.

Mental operations are good when they correspond to the objective laws of the development of technical systems (remember the analogy with the actions of a helmsman of a ship plying a meandering river). Technical systems develop in the direction of increasing the ideal - this is a rule. When at step 4 the examinee made an attempt to mark the ideal (for the given problem) structure of a ray, this was a correct action. But the supervisor decided that there was a mistake here.

After this is it any wonder that the "purely psychological" approach has brought the inventors practically nothing?

Incidentally, another thing is more important. Applying the operations of the ASIP to Dunker's problem, we were enabled to see clearly the mechanics of the steps: these steps allowed us to reject "empty" variants and led us to an answer by making a detour. Why bump into the wall if you can go round it?

TWO INTERESTING EXAMPLES

Let us return to our problems.

PROBLEM 34

Small plastic cylindrical objects are covered with paint by means of a spray gun. If the spray gun is switched on at full power the cylinders will be covered almost instanteously with a thick layer of paint. The coating is bad and, moreover, takes a long time to dry. If the spray gun is turned to its lowest power the process of applying the paint now takes 30 to 40 seconds and can be controlled: one can easily hit on the right moment when no unpainted parts remain while unwanted layers of paint have not yet formed. However this reduces productivity enormously. In this instance the electrostatic method of painting is ruled out. Mixing additives to the paint is also not permitted. What can be done?

Let us write down the solution from step 2.2 (the terminology is not built in to the specifications of the task).

2.2. The artefact is the cylinder (according to rule 4 we take just one cylinder). The paint (a stream of paint, a spray of dispersed paint) is the instrument (strictly speaking that part of the instrument directly interacting with the artefact). A spraygun without the paint does not interact with the cylinder therefore is not one of the conflicting pair. So this means that our task is to learn how to paint well with a bad (that is to say absent) sprayer.

According to the conditions of the task there can be a lot or a little paint. One has to give preference to the first variant (rule 3). And so, the conflicting pair is the cylinder and a large (too much) quantity of paint.

2.3. 1. A large quantity of paint can be easily and quickly applied to a cylinder (paint can be poured over it or it can be dipped in paint).

2. A large quantity of paint forms an unwanted coat on the cylinder.

The entire problem boils down to removing the superfluity of paint (admittedly it is better not to remove it altogether but only in such a way that the residue returns to the vat). "Common sense" tells you to prevent a superfluous coat from forming: why first apply too much and then remove it? The logic of ASIP is different: a superfluous coat of paint can be applied easily, so all right, let's do it! The artefact is painted (and, moreover, quickly), and it only remains to remove the superfluous paint. In fact the problem "How to apply paint well" has now been replaced by "How to remove paint well".

2.4 Thus the model of the task is as follows: We are given a cylinder and a large quantity of paint, which can easily be applied to a cylinder but in so doing forming a superfluous and unwanted coat.

3.1 Both elements lend themeselves to alteration only with difficulty. The cylinder is an artefact and by the conditions of the problem altering the constituency of the paint is severely limited. Let us use as the alterable element the outside environment.

3.2 The IFS: the outside environment itself eliminates the superfluous coat of paint on the cylinder, although the paint is applied to the cylinder in a large (and superfluous) quantity.

3.3 It is simple to show the cylinder with a thick layer of paint and note what is superfluous (fig. 11,a). Using the MLM method one can show the boundary of the paint (we do not yet know what in this instance forms the "outside environment") in the form of little men (fig. 11,b). In both cases the zone of superflous paint is marked off.

3.4. The solution proceeds further in two directions depending on how we have written down step 3.3.

a) For removing superfluous paint we need some force;

b) This force is unnecessary or even harmful. Why? Obviously since it might remove some of the useful coat along with the superfluous paint ;

for fig. 11,b

a) in order to eliminate superfluous little men some force is needed;

b) but this force is harmful for it can also pull off those little men attached to the surface of the cylinder.

In the drawing we can see at once an important characteristic: the particles of paint are linked together in different ways. The useful little men adhere to the surface and the "superfluous" little men only to each other. The different power of connection signifies that there is one mark by which one can distinguish the "superfluous" little men from the "useful".

Fig. 11

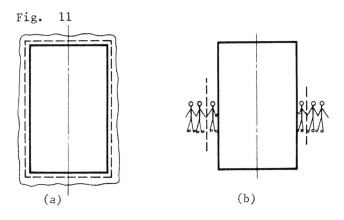

(a)   (b)

By taking into account that two substances are given (the paint and the cylinder) and consequently that a field should be introduced, we are getting near to the solution of the problem. The non-electrical (as specified in the problem) field should tear the "superfluous" (far away from the cylinder) little men and should not tear away the "useful" ones (near to the cylinder).

We do not make use of the MLM since the task is not

complex and the method is capable of "splitting it" halfway
there.

3.5   PC.   a) the delineated zone of the outside
environment   should act on the superfluous paint in order to
remove it, and should not act on the superfluity in such a
way that it takes with it the useful coat.

b) the delineated zone of the outside environment
should and should not exist.

4.1.   Here there is a clear requirement to separate the
contradictory properties spatially. But how?

4.2.   A class 8 problem: two substances mutually
interact, in so doing they are both susceptible to control
(therefore they interact badly). The solution: one must
introduce a field which acts in different ways on these
substances.

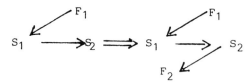

What kind of field precisely? An electrical field is
ruled   out by the specifications of the problem, and
likewise a magnetic field (paint and the cylinder are
non-magnetic and introducing another substance is ruled
out). The field of gravity already exists, but it does not
produce the necessary interaction. There remain two fields,

heat and mechanical. A heat field could spoil the paint, and in a mechanical field paint should be moved and thereby remove the superfluous layer. A mechanical field should be weak at the surface of the cylinder, and strong at more distant layers of paints. Let us examine Appendix 3. For the given problem pp 6, 7 and 12 are suitable. If we examine only the mechanical effects the answer is obvious: centrifugal forces are at work. The cylinder is dipped into the paint and rotated. The centrifugal force hurls the superfluous paint off. The process is controlled by regulating the number of rotations. Many cylinders can be processed at the same time. (Pat. Cert. No. 242 714).

Problem 34 might seem easy. We have cylinders (jars!) and paint - all 19th century level! Would one succeed with the same ease in controlling the process of solving contemporary complex tasks? All right, let's take a contemporary problem then.

PROBLEM 35

For research purposes it is necessary to know the "mobility" of ions in gases 'their speed of directed dislocation). The voltage is given, the distance between electrodes is known, to be measured is the time of drift of the ions from electrode to electrode. This is how one proceeds. At a fixed moment of time, ions are introduced (at the surface of one of the electrodes) and then the time of their drift is measured under the influence of the field to another point (another electrode). To determine the mobility of the ions polarity is switched from one to the other.

On one occasion the requirement was to solve this problem on the condition that the body of gas was being rapidly (30 milliseconds) exchanged. A supplementary requirement was also made: simplicity of equipment. Incidentally, with the increase in the rapidity of action the necessity arose to set up a high voltage start synchronization schedule, devising which calls for considerable time and effort.

Thus in 30 milliseconds one has to measure the extent of drift of the ions of both polarities. If one is to carry out the measurement consecutively then each of them will have to take a maximum of 10 to 12 milliseconds. It is clearly more advantageous to carry out the measurement simultaneously. This is indeed how the problem was phrased. But working according to the DTC operator, carried out prior to the analysis, forced one to return to the principle of consecutive measurement. Mentally we increase the size of the ions: coming to meet each other are oppositely charged tennis balls. We increase their dimensions even more: on collision course move oppositely charged planets. Huge charges unavoidably call up mutual interaction of the planets. But surely such interaction must arise even when oppositely charged ions approach each other! The DTC operator has forced us to pay attention to a circumstance that went unnoticed when the problem was presented. One has to abandon the principle of simultaneous measurement of the mobility of ions of opposing polarities. Let one ion go the distance and instantaneously, without losing time replace it by an ion of reversed polarity. We thus avoid interference, but, alas, we lose out by the undue complication of equipment. one must determine with the very greatest precision the time of arrival at the "finish" of the ion of one polarity, in order to give at once the "start" signal to the next of reversed polarity. Six elements are mentioned in the problem. Two electrodes, two types of ions, gas, an electrical field. What then can be taken as the conflicting pair? The "artefacts" are the ions, the instrument is the field. The electrodes remain "out of play" (just as the paint sprayer in 3 and the chamber in Problem 29).

However, here for the first time we come up against the "electrical specifications" of the problem. Which ion should be included in the conflicting pair, the positive or negative? From the negative ion when it arrives at the "finish" one would "strip off" the superfluous electrons, but "add" the missing electrons to a positive ion at the finish. In effect it is the same situation as in the paint problem: does one move from "too much", removing the superfluity, or from "too little", adding something missing?

When Problem 35 was solved for the first time positive ions were taken... and led to nothing. The negative ions were taken, however, and a new idea was produced. Destroying something (whether it be a house, a statue, a molecule or an atom) is easier than making it; this is unfortunately a rule of life even if for no other reason than that in order to build one needs to bring in material from elsewhere (which is perhaps not in supply) whereas no extra material is needed for destruction.

Increasing entropy is simpler than decreasing it" is another usable formulation.

And so, the conflicting pair is "negative ion - electrical field". The essence of the conflict is that the field is able to drive the ion from one electrode to the other, but at the "finish" the field cannot without outside assistance change the negative field of the ion for a positive. At step 3.1 we take as an object the field (the instrument) and then the IFR would read: "The field itself changes the polarity of the ion while retaining its ability to remove that ion." The PC is clearly seen: "The field should smash a negative ion in order to neutralise it or make it positive, and should not smash the ion to enable it to complete the distance". Just as clearly can be seen the pathway to removing the PC (step 4.1): separation of the contradictory properties in space.

Over the distance the field should not smash the ion but at the finish the field should be different - let it smash the negative ions, let it remove electrons from them (thereby making negative into positive ions). How can this be done? We have already faced a similar task. In a column of air the ions did not appear (no one removed the electrons from the neutral molecules) in weak fields (radio waves) and ionization took place (electrons were removed from neutral molecules) on the appearance of a strong field (lightning).

Thus one needs a heterogeneous field: on the surface of the "finishing" electrode it should have a local reinforcement, a kind of "offshore reef" against which the "ion ship" collides, or a nail against which a rising balloon moves, bursts and sinks down again (already in a different condition than before).

From the point of view of S-Field analysis the model of the task belongs to class 9. A substance and a field interact, exchanging one of these elements is impossible according to the provisions of the problem, (we measure the mobility of ions themselves in an electrical field itself); one must obtain good control of one of the elements. This is a typical S-Field transformation: introducing a second substance lending itself well to control, or transferring the F'l into an F''l field. At the "finishing" electrode there should be a substance which transforms a relatively weak field common to the whole "distance" into a localised strong field.

Thus a positive electrode can have an "igloo" (or in general can consist of an igloo alone). The positive potential should be selected in such a way in order that the tension at the "igloo" is less than the primary tension of the independent main charge (which would set up interference), but greater than the critical tension at which the negative ions disperse. That basically is everything. At the moment of time Tl registered for instance with the aid of an electron oscillograph, on the inner surface of the negative electrode is created a cluster of negative ions which under the influence of the field begin to drift to the positive electrode. On approaching it the ions enter the area of a strong field and there disintegrate into electrons and neutral molecules. The electrons ionise the neutral molecules of gas, causing a flash of the non-independent positive crown in which positive ions are created.

We register the moment in time T2 of the formation of this flash; which at the same time also serves as the indicator of the arrival of negative ions, and the generator of positive ions starting at the moment of time T2 in the opposite direction. Further they register the moment of time T3 of the arrival of positive ions at the negative electrode and obtain in this way three time markers on one oscillogramme, according to which they determine the time of drift of negative and positive ions.

However, this is now a technical detail, it is something else which is important: the solution of such a problem according to ASIP does not differ from the solution of simple problems with jars, the tissue paper or the polishing disk. Only at the very last stage, the transfer from the physical solution to the technical is special knowledge required. One must, for instance, know that the decay of negative ions can be accompanied by a flash. So far this has not been reflected in the table of physical effects and phenomena...

PROBLEMS

The five problems presented below should be solved according to ASIP from step 2.2 to 4.2. Attention should be centred, as before, on the course of the solution, and on accurately carrying out the steps.

PROBLEM 36

An iron ore pulp (a suspension of iron ore in water) is carried along a pipeline. The flow of the pulp is regulated by means of a valve (a damper). But particles of the ore which possess abrasive properties rapidly "eat away" the valve. What can be done?

PROBLEM 37

A method of sealing ampoules. Twenty five ampoules filled with medicine are arranged vertically in nests in a metal container (five rows of five ampoules each). A bank of burners is applied from above (five rows of five burners). Over each ampule is a burner. The flame seals the capillaries of the ampoules. Unfortunately the method has a drawback, in that the flame can only be poorly regulated. It is either too big or too small. Certain ampoules are overheated, certain ones are not completely sealed. One can, of course, turn the flame up full. Then all the ampoules would be sealed but in the majority of cases the overheating would ruin the medicine. One could also apply a very weak flame. Then in no case would the medicine be spoiled but many ampoules would not be sealed. An attempt was made to use a cover, a perforated plastic disk covering the ampoules. However, if the capilleries fit easily into the holes, the flame will pass through. And, if the capillaries can fit into the holes without leaving a gap it is difficult and time consuming to fit the ampoules into such a plastic disk. Moreover this heats up too and transfer heat to the ampoules. What can be done?

PROBLEM 38

In order to demonstrate the rate of acceleration under the influence of the force of gravity a visual aid in the shape of a trolley is rolled down an inclined plane. The trolley holds a dropper - a vessel containing a coloured liquid, secreting individual droplets at regular intervals. A paper strip is laid out along the path of the trolley. If the trolley moves at a regular speed the distance between the drops which have fallen onto the strip will be the same, but if the trolley is accelerating the distance between the points - the drops - will grow.

In order to demonstrate the acceleration as clearly as possible it is necessary to place many drop-markers onto the strip i.e. one needs a very long inclined plane. But it is inconvenient to use a long plane (even if it is portable or folding).

The layout of the apparatus should be retained, but in such a way that more markers are applied to the strip on a small size plane. It is impossible to increase the frequency of secreting drops, and here too we should consider that they fall consecutively.

PROBLEM 39

The chemical application of a coating to

metal surfaces (without electricity) is well known and widely applied. Essentially it consists in placing a metallic object into a vat filled with a heated solution of a metal salt (of nickel, cobalt, palladium, gold, copper). A process of restoration begins, and metal from the solution is deposited onto the surface of the object.

The higher the temperature the quicker the process. But with high temperatures the solution disintegrates, the metal is deposited on the bottom and sides of the vat, the solution quickly loses its effectiveness and in two or three hours it has to be replaced. Up to 75 percent of the chemicals are wasted and this adds to the cost of the process. The use of stabilizing additives does not solve the problem. What can be done?

PROBLEM 40

News item in the periodical The Inventor and Rationaliser (1976, No 6. p. 5): "Good meat sinks - this is the principle underlying the invention (Pat. Cert. No 485 380) of E. G. Savran, H.F.Pankov and V.P.Stoyanov from the All-Union Scientific Research Institute of the Poultrymeat Industry, proposed to judge the quality of meat. The product is consecutively placed into a boiling saline solution of various concentrations. Taking this invention as your prototype make a further improved invention as a follow-up.

5

Forty Basic Methods

IF THE DETECTIVES ONLY KNEW

Probably it is impossible to surprise the reader now with the paradoxical nature of inventions. But here is yet one more paradox: a problem can be difficult only because it is ... simple.

PROBLEM 41

A foreign firm is producing chemical products, surgical spirit in particular, which is transported to various chemical enterprises including a paint factory located five kilometres away from the production factory. Three or four times a week a truck arrives and it is hitched to a filled and sealed tank trailer with a capacity of 10m3 and the truck drives this to the paint factory. There the spirit is pumped out, its volume is carefully measured and the tanker returns to the production factory. For a some time the spirit has been disappearing: each time it is discovered that there is a shortfall of 15 to 20 litres and just before Christmas it has even reached 30 litres. The filling equipment at the production plant and at the receiving factory was checked but all was in order. The tank cylinder was checked - not the slightest crack.

The seals of each tank delivery arriving at the factory were checked - each one was complete. And yet 20 liters were missing once more! Not really such a lot, but still it was annoying and dangerous: for no known reason hundreds of litres were disappearing.

The owner of the firm gave orders for the tanker to be sent under guard, but to no avail. The infuriated boss hired private detectives and these set up observation posts at each point of the route - but nothing helped...

But one day the problem was successfully solved. What, in your opinion, was the answer?

In solving this problem by taking a selection of variants, one usually begins by taking a "new look" at the terms, asking "What if the measuring equipment was inaccurate after all?"..."Perhaps the spirit managed to evaporate from a leak in the tank"..."Did the truck driver do a deal with the guards?"... Then one turns to physics and chemistry: "Could the spirit not have formed a chemical reaction with the substance of which the walls of the tank were made?"... "Perhaps the volume of the spirit has altered because of some change in the atmospheric pressure and the outside temperature?"... However the answer is extremely simple and if the detectives that had been called in had known of the typical methods of eliminating contradictions they would have solved this problem without having to call in the bloodhounds. Method 10: an action which it is difficult to carry out at a given moment in time should be carried out beforehand. It is difficult to snitch spirit from a sealed and guarded tanker, but no problem at all to do it the evening before when the tank is empty and not under guard: go up to the tanker with an empty bucket, no one will stop you... This is what some clever thief did in fact do... The night before a delivery he suspended a bucket inside the empty tank. The next day the tank was filled with spirit, and the bucket filled up too. Then the tank was driven to the recipient plant and the spirit poured off. But the full bucket remained inside the tank. When the empty tank was returned to the manufacturing plant the guard, naturally was called off and the clever thief calmly walked off with his haul.

Attempts to draw up a list of methods have been made for many years. Certain lists contained 20-30 methods. But the selection was carried out subjectively, and methods were included on the lists which for some reason had appeared important to this author or the other. The very concept of "method" itself has no clear definition: the lists can place "splitting" and "analogy" next to each other, although the first belongs to a technical system, and the second to the inventor's thought processes.

Methods used in ASIP are the operators for transforming the original technical system (device) or the original technical process (the method). In so doing, not all transformations, but only those which are powerful enough, in order to eliminate technical contradictions in the solution of contemporary inventive tasks. Such methods can be revealed only by means of analysis of the large mass of patent information, relating (this is vital!) not to all inventive solutions but only to solutions of the highest level (from three upwards).

Work in compiling the list of such methods was begun back in the early stages of the establishment of the theory of the solution of inventive problems. The number of researched authors' patents and certificates grew constantly. The list included in ASIP-71 already contained forty methods. For their elucidation one had to examine the mass of patent information in hundreds of thousands of units and extract over 40,000 strong solutions, which were subjected then to meticulous analysis.

On becoming acquainted with these methods, one should pay attention: many of them include submethods which not rarely form a chain where each subsequent sub-method develops the foregoing one.

You should not be put off by "unserious" titles of certain methods. Of course, in place of "the nesting dolls principle" one could say "the principle of concentrated integration." The essence is the same, but the "nesting dolls principle" can be remembered right off. There is one further consideration: for the sake of clarity and compactness the methods are explained in simple examples, but this does not signify that the methods are fitted only for simple inventions.

THE INSTRUMENTS OF CREATIVITY

Let us examine 40 basic methods of eliminating technical contradictions.

1. The principle of fragmentation
 a) Dividing the object into independent parts.

b) Making the object divisible.

c) Increasing the degree of 'fragmentability' of the object.

Example: A cargo ship is divided into identical sections. If need arises the ship can be made longer or shorter.

2. The principle of removal.

Extracting the "disturbing" part from the object (the "disturbing" property) or, on the other hand, extracting the only necessary part or property.

Unlike the previous method, where it was a case of dividing the object down into identical parts, here the suggestion is to divide the object into different parts.

Example: Usually on small pleasure craft and cabin cruisers electricity for lighting and other needs is provided by a generator driven by the main engine. In order to obtain electrical power while moored one has to install an auxiliary generator with a lead from an internal combustion engine. The engine, naturally, makes noise and sets up vibration. It is proposed to house the engine and generator in a separate capsule placed some distance away from the cabin cruiser and joined to it by a cable.

3. The principle of local quality

a. Switching from a homogeneous structure of the object or environment (the outward effect) to a non-homogeneous.

b. Different parts of the object should carry out different functions.

c. Each part of the object should be placed in the conditions which are most propitious for its operation.

Example: In order to combat dust in mine workings water is applied to the instruments (the working parts of the drilling and loading machinery) in the form of a cone of tiny drops. The finer the droplets the greater the effect in combatting dust, but fine drops form a mist which hinders work. The solution: around the cone of fine drops is a layer of larger ones.

4. The principle of asymmetry

a. Switching from symmetrical to asymmetrical forms of the object.

b. If the object is already asymmetrical increasing the degree of asymmetry.

Example: One side of a car's bumper bar is increased in strength to improve resistance to impact with the kerbstone.

5. The principle of joining

a. Joining homogeneous objects or those destined for contiguous operations.

b. Joining in time homogeneous or contiguous operations.

Example: The dual tandem microscope. One man works the adjuster, while the other is exclusively concerned with observation and note-taking.

6. The principle of universality

The object performs several different functions thanks to which the need for different objects is removed.

Example: The handle of a briefcase also serves as a chest expander (Pat. Cert. No. 187 964).

7. The "nesting" principle

a. One object is contained inside another which in turn is placed inside a third, etc.

b. One object passes through a cavity into another.

Example: "An ultrasonic concentrator of resilient oscillations consisting of interconnected semi-wave sections; in order to decrease the length of the concentrator and increase its durability the semi-wave sections are made in the form of hollow cones placed one inside another" (Pat. Cert. No 186 781). In Pat. Cert.No 462 315 exactly the same solution is used for decreasing the dimensions of the external section of a transformer piezoelement. In a device for drawing metal wire, according to Pat. Cert. No 304 027 the "nest" consists of cones of wires.

8. The principle of counterweight

a. Compensating for the weight of a object by joining it to another which has lifting power.

b. Compensating for the weight of an object by its interaction with the environment (largely at the expense of aero and hydrodynamic forces).

Example: "The centrifugal braking type governor of the number of revolutions of a rotary wind-powered engine placed on the vertical axis of the rotor; in order to maintain the speed of rotation in a small interval of the number of revolutions with a strong increase in the power, the loads on the regulator are made in the form of blades ensuring the aerodynamic braking effect" (Pat. Cert. No. 167 784)

It is interesting to note that the formula of the invention clearly reflects the contradiction overcome by the invention. With a given wind strength and a given mass of loads a definite number of rotations can be obtained. In order to decrease this (if the wind strength is rising) it is necessary to increase the mass of the load. But the loads are rotating and it is difficult to get at them. And so the contradiction is eliminated by giving the loads a form creating an aerodynamic braking effect, i. e., the loads are made in the shape of wings with a negative angle of attack.

The general idea is obvious. If it is necessary to change the mass of a moving body and for some reason the mass cannot be changed, then the body should be given the shape of a wing and by changing the inclination of the wing toward the direction of travel one can obtain extra force applied in the required direction.

9. The Principle of Preliminary Counter-Action

If, according to the specifications of the problem, it is necessary to carry out some action one must carry out a counter-action in advance.

Example: "A method of cutting using a dish cutter rotating on its geometrical axis in the process of cutting; in order to prevent vibrations being set up the dish cutter is charged in advance with forces close in size and direction and directly contrary to the forces arising in the process of cutting (Pat. Cert. No 536 866).

10. The Principle of Preliminary Action

a. Carry out the required action in advance (in full or at least in part).

b. Arrange the objects in such a way that they can go into action without loss of time for delivery and from the most convenient position.

As an example one can take the solution offerred to Problem 41 above.

11. The Principle of the "Previously placed cushion"

Compensate for the relatively low reliability of an object

by accident measures prepared in advance.

Example: "A means of treating inorganic materials, such as fibre glass by means of the action of plasma rays; in order to raise the mechanical stability the inorganic materials are subjected in advance to solution or melting by alkaline salts or alkaline-earth metals." (Pat. Cert. No. 522 150). . Substances are introduced in advance that "heal" the micro-cracks. Or take Pat. Cert. No 456 594 according to which a ring is placed on the branch of a tree (before pruning) and constricting it. The tree, feeling the "pain" sends to that place nutritious and healing substances. In this way these substances accumulate until the branches are pruned, which promotes rapid recovery after pruning.

12. The Principle of Equipotentiality

Changing the conditions of work in such a way that one does not have to raise or lower the object.

Example: A device is proposed which removes the necessity for raising and lowering heavy shape presses. The device takes the form of an attachment with a Rollgang fastened to the press table (Pat. Cert. No. 264 679).

13. The Principle of "the Reverse"

 a. Instead of an action dictated by the specifications of the problem, one implements the opposite effect.

 b. One makes a moving part of the object or the outside environment immoveable and the immoveable part - moveable.

 c. Turning the object "topsy turvey" - reversing it.

On examining Problem 9 (on the dust filter) we got to know Pat. Cert. No 156 133: the filter is made of magnets between which is placed a ferromagnetic powder. Seven years later appeared Pat. Cert. No. 319 325 in which the filter is turned upside down. "The electromagnetic filter for the mechanical cleansing of liquids and gases containing a source of the magnetic field and the filtering element of an aggregate magnetic material; in order to lower the mean consumption of electrical energy and increase the productivity the filtering element is placed around the source of the magnetic field and forms an outwardly closed magnetic profile."

14. The Principle of Spheroidality

 a. Switching from direct linear to indirect linear, from flat surfaces to spherical, from cubical or parallelepipod parts to spherical structures.

 b. Using rollers, ball bearings, spirals.

 c. Switching from direct to rotating movement, using centrifugal force.

Example:A device for welding pipes into a lattice work has electrodes in the form of rotating balls.

15.The Principle of Dynamism

 a. The characteristics of an object (or outward environment) should be changed in such a way as to be optimal at each stage of work.

 b. Divide an object up into parts which can be repositioned in relation to each other.

 c. If an object in its entirety is immovable make it movable, interchangeable.

Example: "A method of automatic arc welding with a strip electrode by which in order to obtain broad control of the shape and dimensions of the weld bath the electrode is bent along its length, lending it a bent shape which changes in the process of welding."(Pat. Cert. No. 258 490).

16. The Principle of Partial or Satiated Action

If it is difficult to obtain 100 percent of the desired effect, one must go for "a modicum less" or a "modicum more" - and the problem may thereby be considerably simplified.

The method is already familiar from Problem 34: the cylinders are painted with a superfluous amount which is then removed.

17. The Principle of Moving to a New Dimension

a. Difficulties involved in moving (or relocating) an object along a line are removed if the object acquires the possibility of moving in two dimensions (i.e. along a plane). Accordingly, problems connected with movement or (relocation) of objects on one plane are removed on switching to three dimensional space.

b. Using a multilayered assembly of objects instead of a single layer.

c. Inclining the object or turning it "on its side".

d. Using optical lines falling onto neighbouring areas or onto the reverse side of the area available.

Method 17a can be linked to 7 and 15c. A chain is formed typifying the overall tendency of the development of technical systems: from a point to a line, then to a plane then to a volume and finally to the joining of many volumes.

Example: A method of keeping the winter supply of logs in the water by placing them side on in water; in order to increase the overall capacity of the equatorial effect and decrease the amount of waterlogged timber, the logs are formed into bundles whose breadth and height exceeds in cross section the length of the logs, after which the bundles are raised to a vertical position. (Pat. Cert. No. 236 318).

18. The use of mechanical vibrations

a. Set the object vibrating.

b. If such a movement is already in effect increase its frequency (going as far as ultrasonic).

c. Use the frequency of resonance.

d. Instead of mechanical use piezovibrators.

e. Use ultrasonic vibrations in conjunction with electromagnetic fields.

Example: "A method of cutting timber without a saw; in order to reduce the exertion needed to insert a tool into timber cutting is effected by an instrument whose pulse frequency is close to the inherent frequency of vibration of the timber to be cut" (Pat. Cert. No. 307 986).

19. The Principle of Periodic Action.

a. Transferring from unbroken action to periodic (impulse).

b. If the action effected is already periodic, change its periodicity.

c. Use the pauses between impulses to bring about a different effect.

Example: "A method of automatic control of the thermic cycle of spot welding, primarily of thin parts, based on measuring the thermo-electro-motive force; in order to raise the precision of control during welding by means of high frequency impulses the thermo-electro-motive force is measured in the pauses between impulses of the welding current" (Pat Cert. No. 336 120).

20. The Principle of Uninterrupted Useful Effect

a. Carry out the work without a break (all parts of the object should be constantly operating at full power).

b. Remove dummy and interstitial runs

Example: "A method for treating orifices in the shape of two intersecting cylinders, for example, nests of ball bearing separators; in order to raise productivity of treatment it is implemented by a drill (Senker) whose cutting edges enable it to cut both in forward and reverse movements of the tool" (Pat. Cert. 262 582).

21. The Principle of Rushing Through

Carry out a process or individual stages (which are harmful or dangerous, for instance) at great speed.

Example: "A means of treating timber in the production of lead by means of heating; in order to preserve the natural timber heating is carried out for a short burst by means of a gas burner with a temperature of 300-600oC directly in the process of preparing lead." (Pat. Cert. No. 338 371).

22. The Principle of "Turning Harm to Good"

a. Using harmful factors, (in particular the environmentally harmful effects) to obtain a positive effect).

b. Remove a harmful factor by dint of adding it to other harmful factors.

c. Strengthen a harmful factor to such an extent that it ceases to be harmful.

Example: "A method for restoring the flow characteristics of frozen aggregate materials; in order to speed up the process of restoring the flow capability of the materials and lower labour consumption the frozen materials are subjected to the effect of extremely low temperatures." (Pat. Cert. No. 409 938).

23. The Feedback Principle

a. Introducing feedback.

b. If there is feedback already, changing it.

Example. "A method of automatic regulation of the temperature pattern of baking sulphide materials in a boiling layer by changing the flow of loaded material in the function of temperature; in order to raise the dynamic precision of supporting a given temperature value the delivery of material is changed in accordance with changes in the content of the sulphurous gas in the exhaust gases" (Pat. Cert. No. 302 382).

24. The "Go Between" Principle

a. Using an intermediary object transferring or transmitting the action.

b. Temporarily joining the object to another (easily removable) object.

Example: "A means of packing devices for measuring dynamic tension in sealed environments during statistical loading of the model environment with a device lodged inside it; in order to raise the precision of packing the loading of the model with the device lodged inside it is passed through a brittle intermediary element." (Pat. Cert. No. 354 135).

25. The Self-Service Principle

a. The object should service itself and carry out supplementary and repair operations.

b. Use should be made of the outlets (of energy, the substance).

Example: In an electric welding gun the welding rod is usually applied by a special device. The suggestion is that to apply the rod a solenoid should be used driven by the welding current.

26. The Copying Principle

a. Instead of an unavailable, complicated, expensive, inconvenient or fragile object a simplified and cheap copy is used.

b. Replace an object or system of objects by their optical copies (depictions). Make use of scale in so doing (scale the copies down or up).

c. If visible optical copies are used, switch to infra-red or ultra-violet copies.

Example: "A visual geodesic teaching aid made in the form of a painting on an artist's panel; in order to obtain a consistent geodesic photograph from the panel on which the locality is depicted, it is made up according to the data of a tacheometrical photograph and at the characteristic points of the locality it is provided with miniature surveyor's poles" (Pat. Cert. No. 86 560).

27. Cheap short life instead of expensive longevity.

Replacing a dear object by a collection of cheap objects, and by so doing abandoning certain characteristics (longevity, for instance).

Example: A one-step mouse-trap. A plastic pipe with a bait; the mouse enters the trap through a cone-shaped opening; the walls of the opening bend closed and do not let it move back.

28. Replacement of a mechanical pattern.

a. Replacing a mechanical by an optical, acoustical or "smell" pattern.

b. Using an electrical, magnetic or electro-magnetic field for interaction with the object.

c. Switching from immovable to movable fields, from fixed to those changing in time, from unstructured to those which possess a definite structure.

d. Using a field in combination with ferromagnetic particles.

Example: "A means of applying metallic coatings to thermoplastic materials by means of contact with metal powder heated to a temperature above the melting point of the thermoplastic materials; in order to raise the durability of the bonding of the coating to its base and its evenness the process is carried out inside an electromagnetic field." (Pat. Cert. No. 445 712).

29. Use of Pneumatic or Hydraulic constructions

Instead of solid parts of the object gaseous or liquid parts are used which are air or water inflatable, an air cushion, hydrostatic and hydroreactive.

Example: In order to link the drive shaft of a ship to the hub of the screw a groove is made in which is placed an elastic hollow container (a narrow "air bag"). If compressed air is introduced into this container it inflates and compresses the hub to the shaft (Pat. Cert. No. 313 741). In such cases a metal linking element is normally used, but a link with an "air-bag" is simpler to make: accurate fitting of the contiguous surfaces is not needed. What is more, such a link smoothes out the impact loads. It is interesting to compare this invention with that published later in Pat. Cert. No. 445 611 for a container for shipping fragile wares (drainage pipes, for instance): in the container is an inflatable envelope which compresses the goods and prevents them breaking in transit. The branches of technology are different but the problems and solutions are absolutely identical. In Pat. Cert. No. 249 583 the inflatable element operates in the claw of a lifting crane. In Pat. Cert. No. 409 875 fragile wares are held by compression in a sawing machine. There is a wealth of such inventions. Obviously it is high time to cease patenting such suggestions, and a simple rule should be written into designer's textbooks to the effect that if one needs to temporarily press one object delicately to another, use an "air cushion". This does not, of course, mean that the whole of Method 29 should be discounted as an invention.

An "air cushion" compressing one part to another is a typical S-Field in which the "cushion" takes the role of a mechanical field. In accordance with the general rule of the development of S-Field systems, one must expect a switch to an F-Field system. Such a switch has really occured: in Pat. Cert. No 534 351 it is proposed that inside the "air cushion" should be a ferro-magnetic powder and in order to step up the pressure a magnetic field should be used. Once more the imperfections of the patent system have led to the fact that it is not the universal idea of the "air cushion" control that has been patented, but the partial improvement of a grinding "air cushion"...

30. Using flexible membranes and fine membranes

a. Instead of customary constructions use flexible membranes and fine membranes.

"A method of forming gas-concrete products by pouring the

mass of raw materials into the mould and the following restraint; in order to raise the degree of swelling a gas-impermeable membrane is placed onto the raw mass poured into the mould (Pat. Cert. No 339 406).

31. Using porous materials

a. Make the object porous or use supplementary porous elements (inserts, covers etc).

b. If the object is already porous fill the pores in advance with some substance.

Example. A system of evaporation cooling of electrical machinery; in order to avoid the need for piping a cooling agent to the machine, the active parts and individual structural elements are filled with porous materials, for instance porous powdered steels soaked in liquid coolant which, when the machine is working evaporate, ensuring short-term intensive and even cooling.(PCN 187 135).

32. The Principle of using paint

a. Change the colour of an object or its surroundings.

b. Change the degree of translucency of an object or the surroundings.

c. In order to observe objects or processes which are difficult to see use coloured additives.

d. If such additives are already used, employ luminescence traces.

Example. US Patent No 3 425 412: a transparent bandage enabling a wound to be inspected without the dressing being removed.

33. The Principle of homogeneity

Objects interacting with a given object should be made of the same material (or one close to it in properties.)

Example: "A method of obtaining a constant casting mould

by forming in it a working cavity according to the standard by the method of pouring; in order to compensate for the settling of the product obtained in this mould the standard and the mould are filled with a material identical to that of the product"(PCN 456 679).

34. The Principle of discarding and regenerating parts.

a. Having fulfilled its assignment or having become unnecessary a part of an object should be discarded (dissolved, evaporated, etc) or changed in shape directly in the process of work.

b. Expended parts of an object should be restored directly in the course of work.

Example: "A method of studying high temperature zones, predominantly in welding processes in which a light-bearing probe is introduced into the inspection zone; in order to improve the possibility of studying high temperature zones in arc and electric slag welding a molten light bearing probe is used which is introduced directly into the inspeciton zone at a speed not less than the speed it melts at"(PCN 433 397).

35. Changing the aggregate state of an object

This includes not only simple changes, for instance from a solid to a fluid but also changes to "pseudostates" ("pseudoliquids") and intermediary states, for instance the use of elastic solid bodies.

Example: FRG Patent No 1 291 210: A braking zone for landing strips made in the form of a "bath" filled with a glutinous liquid, on which is spread a thin layer of elastic material.

36. The Use of phase changes

Using phenomena arising in phase changes, for instance,

the use of volume, the separation or absorbtion of a body, etc.

Example: "Muffling or hermetically sealing pipes and taps with different shapes of cross-section; in order to unify and simplify construction it is made in the shape of a glass into which is poured an easily meltable metal alloy which spreads out on hardening and ensures a hermetic bond (PCN 319 806).

37. Application of heat expansion

a. Use expansion (or contraction) of materials by heat.

b. Use various materials with different coefficients of heat expansion.

Example: In Pat. Cert. No 463 423 it is proposed to make the lid of hothouses out of hinged hollow pipes inside which is an easily spreading liquid. With a change in temperature the centre of gravity of the pipes shifts and therefore the pipes of themselves are raised and lowered. Incidentally, this is the solution to Problem 30. Naturally one can also use bi-metallic plates fixed to the roof of the hothouse.

38. Using strong acidifiers.

a. Replace normal by enriched air.

b. Replace enriched air by oxygen.

c. Treat the air or oxygen with ionizing radiation.

d. Use ionized oxygen.

Example: "A method of obtaining a ferrite membrane by means of chemical gas transporting reactions in an oxydising environment; in order to intensify oxidation and increase the homogeneity of the membrane the process is carried out in an ozone environment"(Pat. Cert. No 261

859).

39. Using an inert environment

a. Replace the normal environment by an inert one.

b. Carry out the process in a vacuum.

This method can be considered to be the antipode of the previous one.

Example: "A method of preventing cotton catching fire in the warehouse; in order to raise the reliability of storing cotton it is subjected to treatment by an inert gas while it is being transported to the storage area."

40. Using composite materials

Switching from homogeneous to composite materials.

Example: "An environment for cooling metal during thermal processing; in order to guarantee a given rate of cooling it consists of a suspension of gas in liquid" (Pat. Cert. No 187 060).

HOW TO USE THE METHODS

A selection of methods, like a selection of tools forms a system, whose value exceeds the arithmetical sum of the values comprising the total of tools. But of themselves in certain instances the individual methods give excellent results. Interesting in this regard is the study carried out by the Sverdlovsk inventor, Candidate of Technical Sciences V.E.Shcherbakov. In technology there is a rather well applied heat-mass exchange apparatus called the Venturi pipe (the accelerator washer, the Venturi scrubber, the turbulence washer). This is a simple pipe contracted in its middle. The speed of the gas passed through it at the place of contraction increases, the gas divides the liquid introduced into the pipe and mixes up with its particles. In essence this is a simple pulveriser. But the pulveriser works with a small volume of substances and the Venturi pipe sometimes has to reckon with a throughput of tens of thousands of cubic meters of gas an hour. As throughput increases so the dimensions of the apparatus grows unacceptably. As the name itself shows, the apparatus has a shape extended in length, therefore it can be regarded as an object with a linear composition. According to method 17 such objects should develop in the direction "line - plane - volume". On this basis V.E. Shcherbakov made a number of compact and powerful thermal mass apparatuses with plane and volume components (Pat. Cert. No 486 645 and others).

A good knowledge of methods noticeably increases the creative potential of an inventor. Therefore in Bulgaria a list of methods forming ASIP-71 has been published as a separate book. Each method is illustrated by many examples, enabling one to better feel its possibilities.

At the same time as explaining the methods they were compiling and gradually improving the tables of applications of methods for removal of typical technical contradictions (13). In the tables are written indicators which it is necessary to change (improve, increase, decrease) and also the indicators which are cannot be allowed to deteriorate if one uses normal (already known) methods. The methods are noted in the table boxes at the intersection of lines and columns. In the latest modification of the table there are 39 lines and 39 columns. Not all boxes are filled in but even so the table indicates the methods for more than 1200 types of technical contradiction.

In compiling the table for each box one has to define the most advanced branch of technology in which the given type of contradiction is removed by stronger and methods with greater long term prospects of success. Thus, for contradictions of the "weight - continuity - action" "weight - speed", "weight - stability", "weight - reliability" varieties the most suitable methods are to be found in inventions coming from aviation and space technology. Contradictions associated with the need for increasing accuracy, are most effectively eliminated by methods inherent to inventions in the field of equipment for physics experiments.

The table of application of methods used in leading branches of technology helps one to find strong solutions for ordinary inventive problems. In order that the table be also suitable for problems which have only just arisen in leading branches it should in addition contain the very latest methods which are beginning to enter inventive practice. These methods are most often met with not in "successful" inventions for which author's copyrights have been issued, but in applications rejected because of "unviability" or "unreality." Thus the table reflects the collective creative experience of several generations of inventors.

For instance, in Problem 28 (measuring existing cones) accuracy of measurement clearly conflicts with its complexity: if you use one certain method you have to operate with a very large number of templates and carry out each measurement with very great care. According to the table (the intersection of line 28 and column 37), we

obtain methods 26, 24, 32, 28. The first method, however
(26), proffers a radical change to a known method. No
templates are needed, and we shall measure not the cone
itself but a copy of it, depictions, photographs.

One must, however, emphasize that the table is by no
means destined for the solution of "raw" problems. The
table is a part of ASIP and should be used in combination
with other of its mechanisms. In ASIP-77 the use of the
table is step 4.4; at first the problem should be
meticulously analysed.

Let us take Problem 28. We are given two substances:
a hollow cone and a template (according to rule 4 the
model of the problem should contain a single pair), there
is no mutual interaction; the task applies to Class 4 as
well as Class 24 on the polishing disk. The solutions are
also similar up to a certain point; in the disk task the
object with a solid composition is replaced by one with a
pseudoliquid (a shifting powder); the template can also be
made pseudoliquid or simply liquid (if there is no
centrifugal force one does not need to think of how to
keep the particles together). Here, however, the
similarity ends, because the disk problem is one of
changing, of processing, whereas the template problem is
one of measurement, of discovering (the gaps between the
template and the cone). Now, when the analysis has given
an idea of the liquid template which can grind easily but
is not capable of measuring, method 26, suggested by the
table acquires a precise sense: one must remove the liquid
template and compare photographs with the control
photograph. The cone is placed in a bath, water is poured
in to a certain level and the level of that located above

this level is fixed with camera. Then water is added to the next level, and again it is photographed on the same plate. As a result the plate shows series of concentric circumferences which can easily be compared with the circumferences on the standard photograph (Pat. Cert. No 180 829).

PROBLEMS

Methods are like instruments - they do not work by themselves. One has to be trained in their use, solve several dozen problems. For a start solve just four problems.

PROBLEM 42
Burning gaseous oil products when moving along pipelines form hard paraffin deposits. One has to shut down the apparatus and remove the paraffin with solvents. It is proposed to satiate the gaseous oil products in advance with solvent steam (PCN 412 230)> What method is used in this invention?

PROBLEM 43
There are rainmaking machines which spray water from a rotating pipe raised above the surface of the ground. The longer the pipe the greater the area which such a machine can inundate. But with the increase in the length of the pipe its weight also increases which in turn complicates the construction of the machine,

increases the expenditure of energy, etc. What method should be employed in order to eliminate this technical contradiction?

PROBLEM 44

As shown in numerous examples the use of the "air cushion" has become axiomatic in the solution of problems in which it is necessary to temporarily compress one fragile object against another. But what if instead of an "air bag" one takes the opposite, a "vacuum bag"? What would it look like? Find an inventive use for a "vacuum bag".

PROBLEM 45

In many technical devices moving bands are used in the form of an endless ring, If, for instance, one covers the outer surface of such a band with an abrasive compound, one obtains a polishing band. In PCN 236 278 a proposal was made to cut through the polishing band, twist one end through 180o and rejoin them, thus obtaining a Moebius loop. Both surfaces of the band have now become abrasive. Its length remains the same, but it is as though it had doubled in length. Othr inventors have done precisely the same with tape recorder tape (PCN 259 499), band filter (PCN 321 266), a band lathe for anode-mechanical cutting (PCN 464 429, nine authors), a conveyor belt (PCN 526 395) and dozens of other belts. What method has been used here?

One further question. The Moebius loop doubles the length of the surface used. But if one takes a triple-sided polishing belt and before joining up into a ring twists one end through 120o then the working surface is doubled (admittedly it has become narrower). One can twist a multifaceted belt and lengthen the surface by five or even ten times. But this invention was patented under PCN 324 137. Forecast inventions which might appear in connection with this copyright.

6

From Simple to Complex Methods

The basic methods and tables of their application are perhaps the simplest thing in ASIP. The application of methods does not require that discipline of thought which is necessary for analysis (S-Field and "in steps"), and demands no knowledge of physics. The table suggests an automatic approach: one does not need to think, one has taken the starting data and received an almost ready made reply. Behind the present small table and short list of methods the optimists see a multitude of larger tables and long lists of methods, and hence with one hand they can be turned over to the computer for application.

After the publication of ASIP-71 many proposals appeared for improving the pool of basic methods. V.D Voronkov, for instance, suggested "reworking" the inventive into organizational methods, destined for solution of general problems of management and organization.[21]. L.C. Gutkin supplemented the list by special (radiotechnical) methods [22]. A.I. Polovinkin divided the methods into a number of sub-methods [23]. Attempts of this nature are undertaken with the very best of intentions but unfortunately are based on purely arbitrary principles. The only way for improvement in the pool of methods is analysis of the broad mass of patent information, relating to inventions of the highest levels.

This road is labour intensive, but it would be worth analysing several hundred thousand inventions in order at the end of the day to obtain "A large table and a long list", if these would guarantee solution of difficult problems. The matter, however, is far more complex.

ASIP-68 included a list of 35 methods, but to obtain this 25,000 patents and copyrights had been analysed. In preparing the ASIP-71 the number of inventions analysed was increased by 15,000 and the list of methods was increased by only a fifth new ones.

Before mechanically continuing the analysis and "scraping the bottom of the barrel" one should take a systematic look at the nature of the 40 methods already published. Which of them are strong and which weak? Why are some methods stronger than others? Is there no way of conducting a purposeful search for new strong methods?

Usually the researchers have identified the strength of a method with the frequency of its application. In fact these are different concepts, and in evaluating the effectiveness of this or that method one must take both factors into account. Such research was undertaken for the first time by D.M. Khiteeva. Having taken a large mass of patent information she first of all weeded out inventions of the first level, and the remaining inventions she divided into 40 types (according to the number of methods), and each type she broke down into three groups: those of the second, third and fourth level. Then for each type (i.e. for each method) she calculated the coefficient of efficiency K according to the formula:

$$K = \frac{a + Lo + Mc}{a + b + c}$$

where a is the number of inventions relating to the first group (2nd level); b is the number of inventions in the second group (3rd level); c is the number of inventions in the third group (4-5th levels); L and M are coefficients characterising the qualitative differences between inventions of the second and third group and those of the first.

If one is to take the small values L and M which are not very different from each other, (3 and 5, for instance), then K is basically considered the frequency of use of the method. If the values of L and M are great and they differ sharply from each other (10 and 100 for instance), the calculated efficiency will practically depend only on the number of inventions of the third group. Therefore D.M. Khiteeva took L = 5 and M = 25. In this case the coefficient K could have a value from 1 to 25; if the method produced only inventions of the first group then K = 1; if all the inventions produced by the method belong to the third group then K = 25.

When the values of K were calculated it emerged that they varied within very broad boundaries: from 3.9 (method 3, the principle of local quality) to 21.3 (method 34, the principle of discarding and regenerating parts of objects).

By juxtaposing strong and weak objects D.M. Khiteeva came to interesting conclusions. It emerged that weak methods are old and directed at the specialization of objects, whereas strong methods are considerably newer and directed at bringing the object close to the ideal machine method or substance. In strong methods are implemented principally new (reverse) approaches (methods 13 and 22), use physical effects (methods 28 and 36), the changes are more delicate and "clever" (method 16) than the older and weaker methods. Let us examine, for instance, methods 19 (the switch to intermittent action) and 20 (the switch to continuous action). At first glance the methods are related. But with method 20 the coefficient of efficiency turned out to be 50 percent greater than in method 19. Why? The continuity of action is approaching an ideal approach, and intermittency is moving away from from it, and this method is justified only in those special cases in which the transfer to an impulse regime leads to a new effect somehow making up for the losses in time during the pauses.

Method 9 (preliminary counter-action) turned out to be stronger than the "related" method 10 (preliminary action). The point is that method 9 in reality includes two operations: making something in advance (method 10) and doing the opposite (method 13). The "dual" method naturally leads to a more radical transformation of the object and is therefore stronger than the singular method.

And so, strong methods:

   - offer radical changes in the object
   - are directed at taking the object nearer to an ideal machine;
   - are a synthesis of several actions.

All these requirements are met simultaneously by sub-method 28: the use of a ferromagnetic powder and a magnetic field (i.e. replacing a mechanical by an F-Field system). It was interesting to calculate the coefficient of efficiency for the F-Field inventions. It turned out to be high - 23.7.

METHODS FORM THE SYSTEM

Imagine that the world consisted only of chemical elements and their isotopes. Possibly it would only comprise several hundred simple substances. The real world is immeasurably richer and this richness has occured thanks to the fact that chemical elements enter into combinations, form complex substances (or to be more accurate many classes of ever more complex substances).

So it is with methods. Like chemical elements they are first of all rarely encountered in the pure form. Let us examine, for instance, the following example of Method 1. The ship is broken down into segments. Is this the principle of splitting? But surely one can consider this to be method 5 - the principle of joining together. The segments are joined together in the ship's hull. In fact both methods are used here: first the hull is divided into segments (splitting) and then these segments are united in a single structure (joining together). The effect is achieved notably by the common application of two methods: direct and reverse.

As I.M. Flikshteyn has shown, all methods can form pairs, method - counter method. Certain of the forty methods are precisely such pairs (discarding and regenerating parts of objects), others are the "fragments" of pairs, they can be reassembled to make complete pairs. Let us say the principle of local quality (i.e. non-homogeneity) forms a pair with the principle of homogeneity. And even such a "onesided" method as the increase in the number of measurements has a counter-method suitable for the formation of a pair - the use of fine films (i.e. the switch from a volume to a plane).

Physical contradictions as we have already frequently seen, reflect dual requirements: the object must possess a property and its anti-property. For instance, it should be a conductor and a di-electric. The dual "lock" should accord with a dual "key": by their very structure the dual methods are better suited to the elimination of contradictions than single (elementary) methods.

If one is to continue the analogy with chemistry, then one could say the paired methods are the simple molecules O2, N2 and H2. Far more widespread are combinations formed by different molecules. The same thing applies to methods. The stronger the invention the more complex the structure of the "key", the combination of methods utilised in this invention. Let us take Problem 9 (the filter). There had been a filter made up of a many-layered metallic material. It was broken up into tiny particles (Method 1), these particles were joined into a single body (method 5), possessing pores (method 31), which could change in size (method 15) under the effect of an electromagnetic field (method 28). Here we have a whole system of methods; it is sufficient to take one of them away for the problem to be incapable of solution. Difficult problems are difficult precisely because for their solution are needed certain combinations of methods (just as in chemistry H_2SO_3 and H_2SO_4 possess different properties and produce different reactions.

It may be objected that the table of applications of methods only puts forward isolated methods... So now, one must take account of this feature. Let the table suggest that it is necessary to use method 1, splitting. At once one can introduce a correction: first splitting, then unifying the split off parts plus something else, in order to gather these split up parts together as a united whole.

In chemistry there are substances which are of especial significance for the chemical industry - several acids and alkalis and salts. And if one is to develop the analogy with chemistry one can ask whether there is a combination of methods which would play such an important role in invention also. Yes. There are, for instance the S-Fields and F-Fields. The switch from a substance to a full S-Field always includes a common utilization of a group of methods, as we have seen in frequent examples.

There is one further important group of complex methods: combinations into which come the principle of preparatory action (method 10) and the principle of partial fulfillment (method 16). If one is to take just Problem 41 alone (selling the spirit). Its solution shows quite clearly the principle of preparatory action. The spirit is stolen from the tanker in advance, before it has been pumped in. But in fact it is impossible to do this since there is no spirit in the tanker at all! And so in addition one calls on method 16. One has carried out in advance not the whole action, but only a part of it. One has put the bucket in place, which then fills up with spirit.

When the wood bark (Problem 3) is treated with a magnetic compound in order to make it easier to separate the particles of bark from pieces of wood, one has used methods 10 and 16 but in combination with method 28. The bark is accorded in advance receptivity to the operations which will be performed on it later. Incidentally, the use of the combination of methods 10 and 16 has been dubbed the principle of receptivity. In Problem 5 receptivity is assured by the preliminary introduction of a luminescent trace; in Problem 8 by the utilization of a quick melting cladding. This is a typical pathway: one introduces a substance capable later of responding easily to the action of a field.

PROBLEM 46

In the metallic body of a device there is an opening into which a ball has been pressed. After a certain time it is necessary to withdraw the ball but this is difficult, since it has been jammed in tight. It is not allowed to take the construction to pieces. What can be done?

It is hard to extract the ball - it has no receptivity to the extraction effort. One must, before the ball is pressed in, introduce into the opening a substance which later, when it is necessary to withdraw the ball, under the action of a field, implements the pressing effect. "A means of joining parts together one of which is pressed into a deep recess by the other; in order to make it possible to replace the part under pressure, for instance the ball of an indicator tip, before pressing it into the recess one should introduce a drop of water which before the time for extraction is heated to evaporation point under the pressure of steam the ball is forced out. (PCN 475 247).

And so methods and their combination form a multistoried system. On the first level are the elementary methods (splitting up, joining together, the principle of local quality, the principle of asymmetry, etc. There is little future for the idea of building up lists of elementary methods, apart from the fact that these methods are weak. The second storey is the stronger paired methods (pairs of the "method - counter method type"). The third level is the combination of elementary and paired methods with others, i.e., complex methods including the combination of the "receptivity" type, the S-Field, the F-Field.

Methods of the first level are in no way orientated toward technological progress. Is it progress, for example, to increase asymmetry? And perhaps it is more progressive to do the opposite - to increase symmetry (the principle of spheroidality?). Sometimes one is better, sometimes the other. It is impossible to say anything more definite. At the third level appears a clear tendency: the more complex the battery of methods, the clearer it is directed along the line of the development of technical systems. An increase in the degree of responsiveness, a transfer from non S-Field to S-Field systems, the conversions of S-field

to   F-Field   systems   -   these   are   tendencies   of   the
development   of   technical   systems   and   important ones at
that.

The   impertinent   question   arises,   and   what if one were to
go   up   one   further   storey?   There   there must be methods
which   are   not   only   complex   but   always   give   strong
solutions.   To   the   methods   of   the   fourth   storey,   for
example,   one can include the formation of F-Fields, so why
can   one   not seek for other equally effective combinations
of methods?

There   are really similar combinations of methods: they are
not   strong   but   also dedicated: each one is suitable only
for   a   specific class of problem. At the first level there
was   no   such specialisation, but instead the methods there
were   far weaker.

We   shall   return   again   to   the   complex   of   problems
inhabiting   the   fourth   storey.   Now   we   are   faced   with
continuing   the examination of the elementary methods: here
several surprises are still awaiting us.

"MACRO" AND "MICRO" LEVELS OF METHODS

Let us compare two inventions:

In   Pat.   Cert.   No   259 949:   "A simplified construction of
signal   lights,   containing   a   post, a head and a base; in
order   to   rapidly   erect   and   take down the light without
displacing   the base the post is made of component elements
linked   together   by   hinged   joints fixed relative to each
other by hand."

Or Pat. Cert. No 282 342: "The application as a working body for the outlines of a binary cycle of an energy installation chemically reacting substances which dissociate on heating as warmth is absorbed and molecular weight is reduced, and recombine on cooling to the original state."

In both inventions the principle of splitting is used. To be more precise, as we now know, the paired method of "splitting-uniting" (in invention Pat. Cert. No 282 342 this can be seen quite distinctly). The method is the same, but the levels with the invention are different: the collapsible-assemblable post of the signal light is an invention of the first level, whereas the application of collapsible-assemblable molecules" in energy cycles is an invention at least of the fourth category.

Let us examine two further inventions:

Pat. Cert. No 152 842: "A thermal drill for drilling holes; in order to carry out drilling on inclined shafts without halting the process of drilling an impulse burner is joined to the cone by a hinge."

Pat. Cert. No 247 159 "A method of directed drilling of shafts with the aid of artificial deflectors; in order to regulate the angle of entry of the curve of the shaft a polymetallic deflector is used and its temperature is changed."

Both inventions relate to the same technological system and their aim is to obtain the same effect: a rigid construction needs to be given flexibility, the capability of controlled change of the curvature. In the first case method 15 is used (the principle of dynamicism): the rigid construction is broken down into two parts, linked by a hinge; in the second method 37 (heat distribution) the same dynamic process but instead of crude "irons" (hinges) mobility is ensured by extending and contracting the crystal lattice (incidentally here is a typical switch to an S-Field: instead of one substance, two are taken - with different coefficients of heat conductivity, and at the same time control is realised with the help of a heat field). It is exactly the same thing in the first pair of inventions: the very same method (the principle of splitting) is implemented on a macro-level (the collapsible-assemblable signal light) and on a micro-level (the "collapsible-assemblable" molecules).

Each method can be applied on a macro and micro-level. In one instance "irons" are used, and in another molecules, atoms, ions, elementary particles. Any invention has a prototype ("what went before") and therefore four types of operations are mentally conceivable:

1) from macro-object to macro-object (let us provisionally name this the M - M process); for instance the signal light post has been broken down into parts;

2) from macro-object to micro-object (M - m); for instance the invention in Pat. Cert. No 465 502: "A braking device containing a shaft and a brake ring fixed to it under tension linked to the source of driving energy; in order to improve the exploitative properties of the braking ring it is made of piezo-ceramics and as a source of energy a high frequency generator is used". Normal braking systems (as in a car, for instance), work on a macro-level - with the help of blocks, levers, springs, rods, etc. The essence of the invention is the transfer to a microlevel: the brake ring is extended by dint of changes in the

parameters of a crystal lattice;

3) from a micro-object to a micro-object (m - m); for instance "the collapsible-assemblable" molecules instead of normal ones;

4) from micro-object to macro-object (m - M); there are no such inventions: the switch from m - M contradicts the tendencies of development of technology, and demands a "primitivization" of the technical system.

If one is to compare the levels of inventions obtained by means of the first shifts, we obtain the following picture: the shift M - M rarely produces an invention of higher than the third level; the shift M - m as a rule leads to an invention of the fourth and fifth level; the shift m - m usually produces inventions no higher than the third level, if the changes occur within the bounds of a single sub-level (a molecule remains a molecule throughout) and higher than the third level if a replacement of sub-level takes place (a molecule permanently or temporarily is replaced by smaller "units" or a field). Historically technical systems developed in three stages. At first the "newborn" technical system absorbs inventions of the M - M type. Development proceeds slowly without especial shake-ups. Let us say that in a sailing ship (the sail - wind system) the sail is gradually improved. Then a technical revolution takes place: a switch of the M - m type. This at times is seen as the emergence of a new technical system, but in fact the system has shifted from the macro to the micro level. The sails are replaced by the pistons of a steam engine or the paddles of a steam turbine; steam presses on these "sails", the molecules of steam are artifically driven

forward by a heat field. Next come a chain of changes of
the m - m type. The steam engine is replaced by the
internal combusion engine: we have the same "piston-sails"
but the propulsion of the molecules by the "wind" is
carried out in a different way. But in invention Pat.
Cert. No 247 064 the "irons" are finally substituted by an
electro-magnetic field, propelling and dispelling the
ions: "The use of an electro-magnetic pump for pumping
electrolytes in the form of a marine jet engine."
Evidentally a new technical revolution is now inevitable:
the use of fields alone.

So far we have been examining the solution of ready-made
problems. But the reader may well ask, "But how can one
present new problems? After all, that is the most
difficult thing and it is no accident that people say that
a correctly presented problem is half the solution." We
have already seen that strong methods of solution are
strong precisely because they reflect the tendency of
development of technical systems. Therefore methods can
also be used for forecasting the topics of inventions.

Let us examine, for instance, Pat. Cert. No 489 862, "A
device for applying polymer powders, containing a chamber,
a porous barrier, a vibrator and a crowning electrode; in
order to raise the quality of the coating applied to it,
the crowning electrode is made in the form of a ring
equipped with means of displacement, made, for instance,
in the shape of micrometre screws." And so, an electrode,
which previously was fixed is made movable, this situation
can be adjusted by a micrometre screw. One has employed
"irons" - in the M - M type shift. One can with certainty
present a new problem: how can one increase the accuracy

of displacing the electrode (and in so doing automate its displacement)? The answer is apparent: one needs a shift of the M - m type. At once one can point to specific methods: a magneto and electrostriction, a reverse piezo-effect and heat conduction. How reliable is such a prognosis? Such a question can be asked, and how will it be answered? Well, there are other technical systems as well, in which the need appeared long ago to raise the precision of adjustment; one can look at how things have been done in these systems.

In Pat. Cert. No 424 238, for instance: in a device for small installation adjustments the length of the regulating element is changed by heating and cooling; in Pat. Cert. No 409 117 a microinjector with an electrostriction lead; in Pat. Cert. No 259 612 in a device for combining microelements the lead is made "in the form of a film changing its dimensions as a result of heat conduction"; in Pat. Cert. No 275 751 regulating a labyrinth pump is achieved by means of heat conduction; in Pat. Cert. No 410 113 a micromanipulator with a piezo-electrical lead; in Pat. Cert. No 518 219 a device for expelling liquids (i.e. the very same microinjector!) with a magnetostriction lead. There are so many such examples that one can without hesitation include in textbooks the confirmation of the rule: "Remember that the micrometre screw sooner or later will cease guaranteeing the requisite precision and go over to the use of heat distribution, magnetostriction, electrostriction and the reverse piezo-effect. So far this rule is not known: each time someone seeks a solution anew and cries "Eureka!" he is making an announcement that he is disputing with expert opinion.

PHYSICS - THE KEY TO STRONG INVENTIONS

It is not difficult to note that at the macro-level simple combined methods predominate (cut through, turn over, join together, etc.), at the micro-level the body of complex methods almost always include physical effects and phenomena. At the micro-level the world of methods becomes a world of physics and chemistry. Hence the necessity for providing the inventor with information about physical methods, i.e., the inventive possibilities of physical effects and phenomena.

Here two problems arise: how can one proceed in order not to spread knowledge about already known physical effects; how can one augment this knowledge by information "over the whole of physics" and "the whole of chemistry".

"School" (not to speak of "college") physics offers a very powerful and almost universal assemblage of tools. But people usually only do not know how to make use of these tools.

Let us recall just Problem 5. We have a gun, and one need only discover whether a shot was fired from it two days before or not. The problem arose because the event took place before and not at the moment in question. Let us reduce the time to zero (as the operator of the IFS demands). Imagine that in the next room someone has fired a shot (or ten shots, it doesn't matter), then instaneously (in the space of microseconds) cleaned the pistol and handed two guns to you. One must determine which one the shot came from.

There is no particular difficulty in solving such a problem. The pistol which has just been fired has a higher temperature. Hence in general: the solution is that: one should measure physical characteristics, which change in a determined pattern after a shot. However, temperature is a poor indicator since it drops too quickly to the normal value. The shot is accompanied not only by a rise in temperature but also by the impact on the material of the barrel. The barrel is made of steel, steel is ferromagnetic, the natural magnetic field of the earth is magnetizes steel and on being fired a demagnetizing effect takes place. A certain time is needed before for it to be fully magnetized again. In this chain of reasoning the simplest "school" physics is used. But it is enough to find a solution. The "method of establishing the age of a shot according to the forensic ballistic experts by determining the change in time in the physical characteristics of the barrel after a shot; in order to determine the time of the shot fired from a weapon found on the scene of the crime one measures with a magnetic device the degree of magnetization of the barrel and carries out a control firing from the weapon; then one carries out control measurement of the degree of magnetization of the barrel every 24 hours until the moment that the indicators of the device are equal to the degree of magnetization of the barrel at the time the weapon was secured" (Pat. Cert. No 284 303).

Experiments on this problem were conducted for almost six years. The problem on no occasion opened itself to rapid solution by a simple running through of variants. But on being given the hint: "The gun is made of steel," the problem was immediately solved by 30 percent of experimenters. If the hint said, "The gun is made of steel and that is a ferromagnetic material" the problem was at once solved by 80 percent of experimenters (predominantly in the general form: one needed to check how the magnetic characteristics changed after the shot. Without a prompt but with a preparatory treatment using the DTC operator the task was at once solved by 20 percent of testers, who were at the beginning stage of studying ASIP (the first year of study at a Public Institute of Inventive Creativity), and up to 70 percent of more experienced experimenters (in the second year).

The physical effects exist as it were in their own right, and the problem also in its own right; in the thought processes of the inventor there is no reliable bridge linking physics with the inventor's problem; knowledge to a considerable degree is standing idle and is not used.

In problems similar to the "gun" throwing a bridge between the problem and physics is not difficult. Let us formulate the rule (it can be regarded as a consequence of what was said about M - m switches: "If one is dealing with iron (or a material containing iron or one into which iron can be introduced), remember, perhaps, that iron is not wood, water or stone, for each atom of iron has magnetic properties and is very easily susceptible to control - discovery, measurement, changing. In the second half of the 20th century one is ashamed to use steel (but it is very widely used) only as a mass of some inert substance (to put it crudely, like a stick), one must draw into the game the finer ferromagnetic properties of iron."

It is difficult to say how many fine inventions will

appear if the engineers begin to apply this eminently simple rule. Take Pat. Cert. No 518 591: "A maltese ratchet mechanism containing a leading link and the well-known maltese cross; in order to raise the period of service the leading link is provided with sectors of a magnetic soft material with permanent magnets embedded in it, and the maltese cross is provided with films of hysteretic material." The maltese cross ratchet is a very old device. But the material of this mechanism is always used crudely on the macro-level. The mechanism is made of steel, but it is used as if it were wood or stone...

PROBLEM 47

> Given is a spring. Increasing its dimensions and replacing a substance of which it is made (special grade steel) is out of the question. A way is needed permitting one to substantially raise the resistance of the spring without adding to it in any way (not adding any supplementary springs, etc). The method must be eminently simple.

One must suppose that you spotted the solution before reading the problem through to the end. Yes, quite right, the windings of the spring should be magnetised in such a way that an eponymous pole is at its side and upon compression of the spring a supplementary repellant force is set up. Offer this problem to your colleagues (the provisions of the problem should be presented word for word).

Let us take another problem.

PROBLEM 48

The line of electrical transmissions and electrotechnical equipment (for instance a distributor) is openly located at substations, and one must protect it from icing up. With this in mind it has been proposed to lay on the cable and equipment to be protected ferrite layers. Under the effect of the changing current these layers quickly warm up and warm the nearby part of the cable or equipment. But the outer temperature changes: sometimes it is above zero and sometimes lower. What is more in general along the electrical transmission line the temperature depends on a wealth of factors, and can be changing constantly. What can be done? One cannot run the length of the line applying and removing ferrite layers.

Here "school" physics is no longer sufficient. One needs now a physics which is a mite more complex, "college" physics. The IFS: ferrite layers themselves become magnetized in negative temperatures and cease being magnetic when the temperature climbs above zero. The physical effects as an instrument of inventive creativity are all the better in that not rarely they permit the IFS to be implemented literally. There is such an effect (the reader has probably heard of it): upon crossing a certain temperature barrier (the Curie point) magnetic properties disappear, and on the reverse crossing are restored. Consequently, the layer should be made with ferrite whose Curie point is around 0o. If you want the magnetism to "switch itself" on and off then use the transition through the Curie point. There could be a multitude of such examples, but so far inventors most often erect unwieldy and unreliable automatic devices, forgetting that the highest form of regulation is self-regulation. Incidentally here is Pat. Cert. No 266 029: a magnetic sleeve switches itself off and on at a given temperature; and in Pat. Cert. No 471 395: an induction burner has "a crucible made of a material whose Curie point is equal to the given temperature of the applied heat..."

Many people know of the Curie point, but it is less well known that another even more fine effect is associated with this point. If the temperature of a ferromagnetic substance is raised then before passing through the Curie point the magnetic properties of the substance are stepped up. This is the Hopkins Effect. His inventive application suggests itself; in many instances it is advantageous for the working temperature to coincide with that at which the "Hopkins Peak" is observed. Here is Pat. Cert. No 452 055: "A method of raising the sensitivity of measured magnetic amplifiers consisting of using the thermal effect on the core of a magnetic amplifier; in order to lower the level of magnetic noise in the operation of the amplifier an absolute temperature at the core is maintained equal to 0.92 - 0.99 of the Curie temperature of the material of the core."

There is an even finer effect, also associated with the Curie point: the transition through this point is accomplished not "as it comes" (a simple disappearance of magnetic properties, and that is all) but in leaps. Each leap corresponds to a change in the magnetism in a very small volume of the material (10-6 - 10-9cm3). This is the Barkhausen Effect. Here is the application of his invention: according to Pat. Cert. No 504 944 the force on the magnetic material is measured by calculating "the number of leap-like changes in the microstructure."

The rule cited above can now be extended to read: "If you are dealing with steel, use not only its tensile properties, but its magnetic too. If they have already been 'activated' make use of the transfer through the Curie point, the Hopkins and Barkhausen effects."

All right, we have formulated the rule, which includes at least some effects relating to the magnetic properties of substances. But what about the innumerable other (non-magnetic) effects, phenomena, properties?

Evidently one can also formulate several other rules. One of them was cited in the previous paragraph (how to bring about microdisplacements). Yet all the same the rules cover only a small part of the physical effects (whereas we are dealing with a combination of effects!). One needs first of all a table of application of physical effects, reflecting the more typical physical "keys" to the typical inventive problems. Such a table is employed in step 4.3 of ASIP-77. Naturally it can be extended and made more precise. To the table should be appended the "Index of Physical Effects", a reference book which briefly explains the essence of physical effects and contains examples of the inventive exploitation (the

Indexhas   been   worked   out   and   employed   in   exercises conducted   under   TSIP   but   it   has   been   impossible   to   include it   in   this   book).

And   so,   we   have   rules,   tables   and   an   "Index"...   Yet this   is   not   enough.   One   can   compute   tens   of   thousands   of physical   effects,   and   they   should   all   find   their application   in   a   correctly   organised   inventive environment.

It   would   be   good   to   have   some   kind   of   universal   means of   searching   out   the   requisite   physical   effect.   At   first glance   it   seems   less   then   serious   to   present   the   question in   such   a   way.   But   after   all   this   is   the   IFS   and   what would   the   ASIP   be   worth   if   its   principles   could   not   be called   upon   to   improve   the   ASIP   itself.

We   have   already   seen   that   many   problems   can   be "translated"   without   difficulty   into   the   language   of S-Field   analysis.   Let   us   augment   this   language   and   if necessary   enrich   it   -   and   we   shall   translate   all   (or almost   all)   the   tasks.   On   the   other   hand   all   (or   almost all)   the   physical   effects   can   also   be   expressed   using   the terms   "field",   "substance",   "effect".   But   what   if   one   were to   use   S-Field   analysis   as   an   intermediary   language between   the   inventive   problems   and   physics   (chemistry).

We   have   already   seen   how   the   conditions   of   the problem   would   appear   in   S-Field   form.   For   instance:

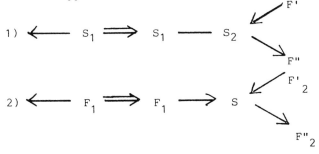

One is given a substance, which is susceptible to control only with great difficulty. In order to control this substance one must link it to the substance S2, which will change its properties depending on changes in the properties of S1. In so doing the changes in the state of S2 should be reflected in the state of the field F interacting with S2, which can easily be ascertained.

Here we are also approaching the physical effect. The ability of S2 to change the state of field F - this is also a kind of physical effect which should be found in order to transfer from the S-Field solution to a physical one.

In the second example a finer effect is required: the substance S introduced ino field F1 should alter its characteristics in such a way in order that this be apparent in the interaction of S and F2.

PROBLEM 49

Measuring super voltages and currents in wires carrying these voltages is a complex technical problem. One has to erect enormous constructions which are fully insulated; such "towers" may be 10 to 12 metres high. One needs to find a simple, cheap and accurate method of measurement.

In S-Field form the solution of this problem has already been mapped out in figure 2. Insofar as acting on the substance is an electrical or magnetic field, it is possible to note this specifically: Fl = Fe or Fl = Fm. At the exit stage it is desirable to have a magnetic, electrical or optical field (other fields are far less suitable). This means one can also specify F2. But then the right hand part of figure 2 gives the formula of the Kerr effect (Fl = Fe, F2 = Fopt) or the Faraday effect (Fl = Fm, F2 = Fopt).

If we had a list of physical effects in S-Field form finding the requIsite effect would present no difficulty, especially since the names of the sought after effects (but not their essence) can be obtained from a general formula linking the names of fields at the entrance and exit (electrooptical, magnetic-optical).

Experienced high school students have found without trouble even more complex physical effects which were known to be unfamiliar to them, naturally in those problems whose solution needed only one physical effect. If a problem is solved by the combined application of several effects (or the combination of an effect and a method) one needs further rules for "joining" physical effects. Such rules are now being studied and certain of them have already been established. For instance, it is known that the "linking" element between two effects to be "joined" in strong inventions is always a field and not a substance (i.e. a field at the exit of one effect is at the same time a field at the entrance to another).

Much remains to be explained. But the general principle is already clear: there is a reliable intermediary between the inventive problems and physical effects necessary for their solution - this is S-Field analysis.

PROBLEMS

If you have been reading the book attentively it would not have been difficult to solve Problems 50 to 52 at once. The following three problems are rather more difficult. First formulate the IFS and the PC for them. Think over what exactly the sought for physical effect should do in order to eliminate the PC. Then use the table of applications of physical effects.

PROBLEM 50

There is a need to automate the process of sorting ripe and unripe tomatoes. Various methods are known (for instance, sorting them by colour, firmness, chemical composition), but these are complicated, expensive and unreliable. We shall take as the basis the simplest (and hence most attractive) method - separating them according to specific gravity. Equipment has been set up consisting mainly of a water-filled vat. The ripe tomatoes should sink in it, and the unripe ones float. Unfortunately this set-up works very badly. Most often ripe and unripe tomatoes have a density of less than 1 gram per cm3... and all will float happily even though the ripe ones are a mite heavier than the unripe ones. It would be most convenient to separate the tomatoes in a liquid with a specific gravity of 0.99 g/cm3. But no such a liquid which would meet the requirements of the food industry has so far been found. Add other liquids to the water, heat it up, pump air into it? What should be done?

Note. It is obviously not difficult to solve this problem. In Chapter 1 an invention was mentioned making use of the requisite physical effect. The very same physical effect is mentioned in the invention cited in Chapter 2 (incidentaly this invention was made by a tenth grade high school student A. Zhdan-Pushkin, a student at the Azerbaijan Public Institute of Inventive Creativity).

It is interesting to compare the answer to Problem 50 with the ideas mooted durng the futile attempts to solve a similar problem by the brain storm (9. pp 60-61).

PROBLEM 51

A problem taken from the "Noticeboard" of the periodical "Inventor and Rationaliser":"Suggest a simple construction of a device for weighing half-railcars carrying scrap metal directly at the place of loading. The permissible rate of error is 1-2 tons. How are the half-railcars weighed at present? Having taken on the scrap metal the railcar is shunted onto the weighbridge by a motor shunter, a process which takes 6 to 12 hours. After this

one has to either add some more or remove some
of the load to bring it the standard amount."

Placing under the rails or the waggon a
tension sensor is a bad solution, and installing
a sensor in the waggon itself is even worse.
What can be done?

PROBLEM 52

Devices are known enabling one to open and
close the way to a gas from vessel A to vessel B
- for instance various taps and valves. But
these are too crude for those instances where a
higher degree of accuracy is needed, i.e., when
one needs to open a tap (or change the extent it
is open) to some extemely small gap.

The tap needs to be very simple and at the
same time work very precisely. This is not a
case for introducing a feedback between the tap
and vessel B. The tap should be controlled by a
human being. The point is making the tap open
accurately ("fauceting").

PROBLEM 53

In a centrifuge chemical reactions are to
take place lasting over a prolonged period. For
this it is necessary to maintain a temperature
in the centrifuge of 250C. It is impossible to
place the centrifuge inside a thermostat (it is
too large). Passing an electric current through
a rapidly rotating centrifuge? Complicated, and
in any case how can one regulate the temperature

within the centrifuge? Should one use heating by infrared rays? The question is once more how to regulate the temperature? After all, measuring the temperature at the surface of the centrifuge is not the same thing. What can be done?

PROBLEM 54.

An extract from a detective novel:

"Now you are in the hands of justice", said the sheriff. "You thought you would get away with it, didn't you? The 'Jupiter' Diamond isn't a bad haul. But we caught you red-handed. The fact that you cut up the diamond into parts and polished it only makes you more guilty."

"Not so quick, sheriff", shrugged one of the prisoners. "So the 'Jupiter' diamond has gone missing? We are deeply sorry, etc, etc., but we haven't got it. We have just got five diamonds. An heirloom from granny."

"That's it," said the second prisoner. "Look at it scientifically. The weight is different, the shape is different. Is the light the same? There are plenty of white diamonds and brilliants. Chemical composition? You've got carbon? We got carbon! All diamonds got carbon. Maybe you should just let us go, what do you think?"

You are all familiar with this situation. What do you suggest?

PROBLEM 55

In Gordon's book Synectics there is a passage taken from notes on the solution of a problem on the transmission of rotation. The text of the notes are also reproduced in V. Orlov's article "The Fireworks of Discovery" (Technology and Youth, 1973, No 3, p. 4). The transmission shaft develops from 400 to 4000 rpm but the drive shaft should rotate at a constant 400 rpm. How can this be maintained? In the notes taken by Gordon empathy is used. One of those tackling the problem mentally put himself inside the "black box" (the device under investigation); in his hands he holds the transmission shaft, in his feet the drive shaft; in this process the efforts of the "empathiser" are directed at ensuring that "in his feet it turns at 400 rpm however much his hands rotate. The answer is not cited.

What are your suggestion? On what rules are they based?

7

The Strategy of Invention:
Controlling the Presentation of Problems

"THE LIFELINE" OF TECHNOLOGICAL SYSTEMS

The life of a technological system (as incidentally in other systems, including biological) can be presented in the shape of an S-curve (figure 12), showing how the main characteristics of a system change in time (power, productivity, speed, the number of systems produced, etc).

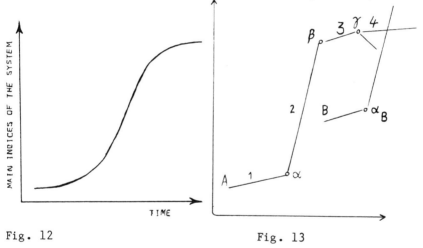

Fig. 12   Fig. 13

In different technical systems this curve naturally has its own individual features but it always has characteristic segments, which are brought out schematically with crudely depicted emphasis in figure 13.

In "childhood" (segment 1) the technical system develops slowly. Then the time of "growing up" and "maturity" (segment 2) comes - the technical system is quickly improved, and its mass application begins. At some point the rates of development start to fall (segment 3) and "old age" sets in. Later two variants (after point γ are possible. Technical system A either degrades,, changing into B, a system that is different in its principle, (modern sailing ships do not have the speeds attained by the famous tea clippers of one hundred years ago), or for a long time stick to achieved values (the bicycle has undergone no essential changes in the past half-century and has not been displaced by the motorcycle.)

What does the relationship between the segments depend on? In other words, what determines the position of the points α β and γ) of curve on the "life graph" of this or that technical system?

Study of the curves of the development of the parameters of different technical systems (the speed of movement of airplanes and ships, the speed of drilling, the growth of the energy of accelerators, etc) forces one at once to pay attention to the fact that the real curves differ noticeably from those expected in theory. The character of the difference is shown in fig. 14 where the dotted curve is theoretical and the completed one is real.

It would seem that from the moment of its appearance a technical system should unswervingly (although not very quickly) develop up to α i.e., until the moment of its transition to mass application. In fact the transition to

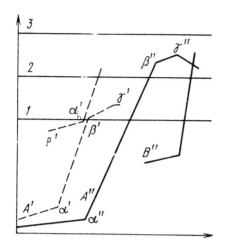

Fig. 14

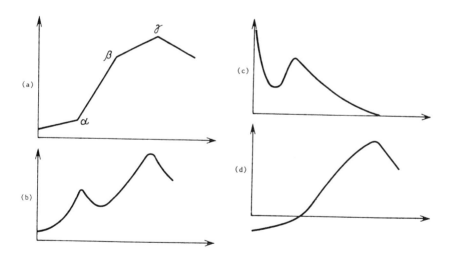

Fig. 15

mass application (a") begins after a delay and at a lower technical level.

The period of rapid development of technical systems should be completed in point β, where the possibilities for the utilization in the system of the principle are exhausted and the economic unviability for the future development of a given system (level 1). However, nothing of the kind occurs; the real point β " is always far higher than the theoretical β'. When the curve A" reaches level 1 many people turn out to be interested in the future development of the system An momentum of interests gets going - financial, scientific (pseudoscientific), careeristic, and simply human (fear of abandoning an accustomed and time worn system). One can ask, does this mean that the momentum of interests turns out to be stronger than economic factors? Yes, it is stronger. But the economic factors themselves are able to adapt to the momentum of interests. Right down to level 2 the system continues to remain advantageous at a cost of destruction, pollution and chemical exploitation of the environment. A typical example is served by the accelerated construction of large tankers in capitalist countries. The famous Torrey Canyon disaster (where 120,000 tons of oil were spilled into the sea) led to severe consequences for the British and French coastlines. Since that time the oceans have not become calmer and sailing the seas has not become safer. But already tankers are being constructed of half a million tons and tankers are on the drawing boards with a water displacement of one millions tons. The curve A" is reaching level 2. Economics (i.e. profitability for the ship owners) is ensured at the expense of damage to the environment. The number of large tankers is increasing, their speed is also increasing (although so far no effective solution has been found to the problem of braking), and inexorably the danger of a super-catastrophe is growing.

"This is good for me today, and I don't give damn for the rest", is the formula that pulls curve A" upwards, to level 2 (economical at the expense of causing harm to the environment). And then all the same a ceiling is reached - level 3, determined by physical limitations. One cannot, for instance, squeeze onto the roads more cars than they can take when cars are already standing bumper to bumper.

Theoretically, as long as curve A' rises to level 1 someone ought to develop technical system B' in such a way so that its point of lift α'_b coincides with point β' of curve A' and ensured a constant unstepped rise. In fact the real curve B" begins to perceptibly rise only when curve A" has risen above level 2 and approached level 3 (for instance: work on a "clean" car). But a rapid rise of curve B" occurs only after curve A" passes point γ " and starts falling.

In fig. 15.a the familiar "life curve" of a technical system is depicted. It is interesting to compare this graph with those characterising purely inventive indicators.

Fig. 15.b shows a typical curve of change of the number of inventions relating to a given technical system. The first peak corresponds to point α (fig. 15.a): the number of inventions increases in the period of the transition to mass application of the system. The second peak in fig. 15.b is conditioned by the striving to prolong the life of the system.

The change in the level of invention is indicated in fig. 15.c. The first inventions setting the basis of the technical system are always of a high level. Gradually this level is lowered. The peak on the drawing corresponds to inventions which ensure the system the possibility of mass utilization. Beyond this peak comes a fall: the level of inventions inevitably falls, approaching zero. And meanwhile new inventions of a high level are appearing relating to system B.

Finally in fig. 15.d is shown the change in the mean efficiency (the practical produce, the economies, the "benefit") from one invention in different periods of the development of a technical system. The first inventions despite their very high level do not bring in profits; the technical system exists on paper or in isolated prototypes, it contains many minor faults and imperfections. Profit begins to appear after the transition to mass application. In this period even a small improvement brings a large "economy" and accordingly a greater reward for the authors.

The inventor needs to know the characteristics of the "life curve" of technical systems. This is necessary for a correct answer to a problem of vital importance for inventive practice: "Ought one to solve a given problem and improve the technical system specified in it or present a new problem and arrive at something which is fundamentally new?" In order to answer this question (step 1.3 in the ASIP) one has to know what are the development reserves of the given technical system.

One can almost always collect information about the course of the preceding development and construct a graph of change of one of the main indicators of the system (speed, productivity, power, accuracy, etc.). Three cases are possible here:

1. The technical system has not yet reached point α. The problem consists in discovering this point. A typical error is attempting to forecast this point by extrapolation from the possibilities of development of the given technical system. In fact point a for the given technical system does not arrive before the preceding technical system has begun to "die out", whose existence is restraining the growth of the young "rival". For instance the "postautomobile" (i.e. a technical system to replace the car) will be intensively developed only when the development of the car reaches its physical limit (fig. 14, level 3). If today one hundredth of the resources and effort were to be put into the development of an electric car as has been put into the development of the present car, the electro-car would rapidly reach point a. But this will not happen. The car can still develop between the first and second stages and will develop even though its use is polluting the atmosphere.

And so, in forecasting the development of a technical system at the initial stage (up to point α) one should get one's orientation from the state of the preceding technical system.

2. The technical system has passed through point α but not yet reached point β. In this instance forecasting consists in defining the second and third stages. In the extreme instance it is sufficient to define only the third

stage, because there exists a clearly marked (although undesirable) tendency toward reducing the distance between the second and third levels. Definition of the physical limits usually presents no special difficulties; they are tied to objective factors lying in full view (such as the durability of properties of the material, the calorific value of fuel, the different barriers - sound, heat, etc.).

3. The technical system has passed through point β (or γ). In this situation the forecast amounts to seeking out a new technical system to which the "baton" should pass.

In each of these instances the inventor has to work in two ways. Let us suppose that the technical system has not reached point α. The inventor can concern himself with improving the "newborn" technical system. This promises major inventions: from a New Idea one should make a New Thing, and for this Idea (an invention of the fifth level) one must accumulate inventions of the fourth and third levels. But the path to point a may turn out to be long, very long even; this, as we have seen depends on the vital resources preceding the technical system. It cannot be ruled out that the waiting period exceeds a lifetime. On the other hand the possible gains are great. The inventor's fame goes mainly to the one who has made a New Thing of practical advantage. Those who have spoken of (and even taken out a patent on) a New Idea are mentioned much later.

Which route to choose - should one set about creating a New Thing or occupy oneself with small improvements to another already acknowledged technical system (preceding point $C^{\prime}$) i.e. which is better - a crane in the sky or a tomtit in the hand - this question goes beyond the limits of the rhery of the solution of inventive problems. Theory can only demand that the inventor sees the two possibilities and consciously picks one. The choice, however, will depend on the attitude to the world and what the individual considers most valuable for himself.

The problem of choice remains whenever the technical system is passing through a stormy period of development in segment α to β . For development in this segment the system needs preeminently inventions of the second level, but in great quantity. Almost guaranteed is success and the possibility of obtaining dozens of patent certificates, the comparative simplicity of implementation - it is no easy matter to surrender all this up and give preference to the next technical system which is pointing forward into the unknown. But even here people not rarely go against "common sense". The specialist in steam turbines suddenly abandons the Thing and goes over entirely to gas turbines which exist in the form of a dubious Idea...

The most striking thing is that the problem of choice is preserved even in those instances when the technical system is known to be old and even decrepit. Here there is no hope of any kind of noticeabe creative success: an obsolete system assimilates only inventions of the first level. But after all one can assemble hundreds of patents, and, which is the main thing, it can be done in all calmness without the torments of creativity and other experiences.

One hundred years ago drum furnaces appeared. These

were small (measuring a few metres) cylinders; raw materials were introduced into them at one end and hot gases from an oil burner at the other; the furnace rotated slowly, mixing the raw materials. The modern cement furnace is of gigantic construction: in length it measures from 200 to 250 meters, in diameter it reaches 5 to 7 metres, and supergigantic projects it can reach a length of 350 to 400 metres and a diameter of 8 - 9 metres. Along this colossal rotating tunnel moves a small streamlet of burning raw material. In order to transfer to this streamlet the heat from the flow of gases, chains are suspended in the furnace - many chains weighing 100, 150, 200 tons... The bigger the chains the better the transfer of heat, but the heavier the furnace the greater the formation of dust from the chains pulverising the raw material. And so we have a flow of inventions on the theme of "Let us hang the chains in a different way": Pat. Cert. No 226 453 (some chains are hung under others), Pat. Cert. No 187 606 (chains are suspended like a spiders' web), Pat. Cert. No 260 483 (again a spiders' web but in another configuration), Pat. Cert. No 266 484 (another drawing of a spiders' web) Pat. Cert. No 310 095 (yet another web), Pat. Cert. No 339 743 (the link of the chain has not two "irons" but three, a capacious "pretzel").

I was interested in the psychology of the specialists who have come up with such inventions. In answer to my questions one of them said, "You see, I work in an institute, and it has a plan, the department has a plan too, my group has a plan. No one writes in the plan that by such and such a date you have to invent a fundamentally new method of obtaining cement. In the plan they write,

"By such and such a date work out such and such an improvement to such a such an entanglement..." I answered, "We have to fulfill the plan. What is preventing you from finding something fundamentally new apart from the plan? For instance, no one forced A.G.Presnyakov to invent an MGD-driven ship." He shook his shoulders: "You are talking about people who took risks and won. But after all not everyone who aimed high has achieved success. And how many years do you have to wait for that success?"

In this book I have decided not to bring examples from my own inventive practice: the theory of invention should be built on objective data and not on isolated chapters. But nevertheless there is one example I shall cite.

In 1949 an All-Union competition was announced for a heat-resistant suit for mining rescue teams. The specifications were there the suit had to protect a man for two hours in an outside temperature of 100o and a relative humidity of 100 percent, and the whole costume was not to exceed 8 to 10 kg in weight. The task was regarded as insoluble in principle. Even using the most modern cooling agents the weight of the suit would come to over 20 kg. It is out of the question to load a mining rescue man with 28 to 30kg when he was already carrying a breathing apparatus (12 kg) and instruments (7 kg).

One could take the problem as it was presented by the competition organizers. In the final analysis if you make a suit with a small store of ice and a reflecting surface it would not be difficult to arrive at 8kg. Of course this suit would give protection for around 15 to 20 minutes, no more. But all the same this is better than nothing. There

was   another   route: change   the problem and let us create
not   a refrigerating suit but another technical system, let
us   go to the outlet side. This is the route that we (I was
working with R.B.Shapiro) chose.

We   solved   the   problem in this way: we threw out the
breathing   apparatus and gained 12 kg, we added this to the
10kg   released for the refrigerating suit, we calculated on
a gas-heat   protective   diving   suit   working on a unitary
refrigerating   substance: liquid   oxygen   evaporates   and
warms   up,   absorbing   the   heat,   and   then   is   used   for
breathing.   We   obtained   an   enormous   reserve   of
refrigerating   capacity   (one   could   work for an hour in a
furnace at 500 C.) and a handy breathing set up.

The   result   was three variants of a diving suit - and
three   prizes   in   the   competition. Twenty years later the
cover   of   the magazine Sovetskiy Soyuz carried a colourful
picture: gleaming   in   reflected   flames   an   experimental
model   of   a   gas-heat   protective suit. This was already a
Thing,   and   I   examined this photograph without in any way
regretting   that 20   years   before   it   would   have   been
possible to move along a simpler route.

THE ROUTE TO THE EXIT

Breathing   apparatuses   have been in existence for 100
years.   Even   a   very talented designer could hardly reduce
its   weight   (without cutting down on other qualities) even
by 0.5   to   1kg. But   we   made   the transition to another
technical   system   in   which the breathing apparatus became

only a sub-system and gained 12 kg; breathing now was ensured incidentally to a new main function - defence against heat. It was a psychological barrier that had to be overcome.

Similar barriers arose when the technical system passed point β. The inevitability of change of the system became evident, but the limit of development of the given system was perceived as the limit to development in general. One forecast that it would be impossible to give up the customary technical system.

For instance, in the 1930s in the developed countries the number of cinemas per head of population grew rapidly. But it seemed completely obvious that a fall would come long before there was one cinema per head of population. In fact this did not happen: television came ("a cinema for every member of the population").

One should pay attention to one extraordinarily important feature: television is not just cinema but it also depicts events, is a platform for speeches, "a newspaper in pictures", etc. TV became the next "step" after cinema having taken it into itself as a subsystem. The simple home cinema (using a cinema projector) did not receive genuinely mass distribution and never reached point a.

System B comes in to replace System A, including it as one of its subsystems, - this method uses System B in order to overcome the oppressive effect of system A and blocking the influence of the inertia of interests.

A witty method of overcoming a contradiction: system A is retained and yet not retained.

The car may possibly be replaced not by an

electro-car but by a system that includes a car (or a means of transport equivalent to it) as one of its subsystems.

It is curious that in forecasting this law has not yet been recognised. Examining, for instance, the curved growth of the production of normal refrigerators, the forecasters discuss the notion that "satiation point is bound to come", and that "there cannot be 10 refrigerators for every person." In fact there will be and will not be refrigerators - they will come in as subsystems in a more universal technical system (a conglomerate of air conditioner, refrigerator, stove etc.): evaluated as conventional utilities this would amount to 10 refrigerators per person.

The law stating that "A technical system is raised to a qualitatively new level becoming a subsystem of a more general level" is of extraordinary importance for an understanding of the mechanics of development of technical systems. In order to correctly apply this law in forecasting the development of technical systems it is necessary to keep it firmly in mind that development is irresistible: a technical system will develop whatever "impossibles" are raised, but in another (perhaps unrecognisable) form (having become the subsystem of another system).

Here one often encounters strong psychological hurdles.

At a seminar on the theory of the solution of inventive problems the students were given a piece of homework: to forecast (naturally in only the broadest form) the further development of a tanker fleet whose curve of development (the growth of global tonnage) is now located somewhere between points β and γ . Beforehand they were told (without undue emphasis) about the law stating "going upwards means becoming a subsystem". However, in their work at home no one made use of this law. The hypnotic effect of the "obvious" was too powerful. All their works proceeded from the fact that the present high rates of growth of the global tonnage of the tanker fleet cannot be maintained for long. Keeping up such rates would lead in 20 or 30 years to there being more tankers than all other ships (including tankers) put together. This is clearly impossible since a part cannot be greater than the whole.

Proceeding from this "obvious fact" the forecasts were produced. Ideas were mooted such as "there will be fewer, faster tankers", or "the demand for oil will drop","oil pipelines will be used instead of tankers", etc.

The main advantage of the tanker is that it is a cheap cargo carrier. Therefore the tanker does not need the speed which would result in a rise in the cost of transport. One cannot expect in the next few decades any fall in the demand for oil. For a long time oil has been not only a fuel but also a raw material for the chemical industry. There is no basis for believing that transoceanic oil pipelines (if they were to be successfully built) would turn out to be more reliable or safe than tankers. Experience has so far indicated the reverse. In 1974, for instance, a submarine pipeline belonging to Shell ruptured. For days on end the surface of a river was covered with oil for a distance of around 140km. Ninety kilometers away from the accident flocks of water birds were found killed by the oil.

"There should be many tankers but fewer of them" - here we are dealing with a contradiction and one cannot merely look at one half of it ("there ought to be fewer tankers therefore and so they will be displaced by oil pipelines"). The correct solution always satisfies both parts of a contradiction.

Carried to its final conclusion "many" means "the whole fleet" and "few" means "not a single ship". The whole fleet and not one ship ... i.e. the whole fleet if necessity arises should be able to convert into tankers (oil carrying) and once more become non-tankers. This removes the contradiction. And this accords with the law of "in developing it becomes part of a subsystem."

Technically there are many ways of making a "multipurpose" fleet. One of these is to build segmented ships in which a small engine section (the "locomotive") is coupled to the large cargo section (the "rolling stock"). The cargo sections can be larger than the engine sections (railcars are larger than locomotives): one implements thereby the apparently improbable situation where "there are more tankers than the entire fleet."

Another route is also possible. Building ships capable of transporting any cargo in standard containers. Here is one of many reports: "The Norwegian engineer V. Foneland has applied for a patent for a new design of freighter for transporting fluid cargoes placed in cylindrical containers of great capacity (5000 cubic metres). The architecture of a ship designed in this way reminds one of a tanker and has astern large watertight gates for loading containers" (Morskoy flot, 1974, No 12. p.52). In the containers can be any liquid or pourable cargo: this is not an (oil) tanker and yet it is.

Having been given a problem the inventor should determine whether to solve the problem itself or to make a detour (steps 1.1 - 1.3 in ASIP-77). There are two kinds of criterion to be applied here: objective (studying the "life graph" of the system) and subjective (one's personal disposition toward a "large" or a "small" invention). Practically when searching for detour routes it is convenient to make use of a system operator (step 1.2).

The sense of the system operator is that the problem is changed by making a shift to a supersystem or a subsystem, and at each level - by a shift to an anti-problem, the reverse of the given problem.

Let us take, for instance, Problem 29 - on increasing productivity in making the drawings for a cartoon film. According to the specifications of the problem the system consists of a drawing (or a series of drawings). At system level the problem reads: how quickly can one move on from drawing A to drawing B? At subsystem level the problem is drastically altered: there is a particle of a substance (a drop of paint), "pieces" of the stroke applied by pencil, in short, there is some small quantity of a substance from which a drawing is formed; how can one control the transfer of this substance? Presented in this way the solution is very easy. We have already had Problem 3 on the transfer of pieces of bark. One has to add a ferromagnetic substance to the substance and use this for the transfer of a magnetic field. One does not need to redraw each time a new position of a line but merely move the same line, changing its shape. "A method of reproducing a silhouette for filming in cartoon films; in order to reduce the labour time required in the process the outline of the object is formed by applying it to a magnetic panel filled with a cord of ferromagnetic powder and a change in the silhouette by relocating the object relative to the point of view is obtained by moving the cord over the panel."(Pat. Cert. No 234 862).

Changing the position and shape of the whole picture is difficult; the mind literally boggles at such a task. Changing the position of the particles of the substance is easy, especially since this has already been encountered in other problems. The hurdle is purely psychological, but extraordinarily high... if one does not use a system operator.

Very interesting transformations occur with Problem 29 in switching to the super system level (the drawing is only a part of a more complex system, even including the cine camera and means of illumination; (can one make an animated film by filming, for instance, an immovable puppet and adding movement only by dynamic movement of the camera and lights?). No less interesting is the transformation of the problem into an anti-problem. In Problem 29 (at system level) one needs to apply black lines on a white background but in the anti-problem one removes what is not wanted from a completely black background, leaving only the necessary lines.

The system operator is not designed for the solution of problems, although sometimes transformation of the problem automatically leads to its solution. Designating a system operator to help in the selection out of a detour problem which can then be solved by ASIP from step 1.4 etc. Like the DTC operator the system operator is a powerful tool for training the imagination.

LAWS OF DEVELOPMENT OF SYSTEMS

The law of "developing by becoming a subsystem" is of obviously fundamental significance not only in technology but also in the development of an object - from elementary particles to galaxies. However, this is but one of the laws which the inventor has to know.

Laws of development of technical systems can be divided into three groups: "statics", "cinematics" and "dynamics." Let us begin with "statics", the laws which determine the start of life for technical systems.

Any technical system arises as a result of a synthesis of various parts into a single whole. Not every combination of parts can produce a system capable of life. There exist at least three laws whose fulfillment is necessary for the system to be capable of life.

1. The law of the completeness of parts of the system
A necessary condition for the living capability in principle of technical systems is the presence and minimal functioning power of the basic parts of the system.

Each technical system must include four basic parts: an engine, a transmission, a working organ and an organ of steering. The meaning of Law 1 is that for a synthesis of a technical system four parts have to be present as well as their minimal suitability for carrying out the functions of a system, for of itself the functioning capability of part of a system can prove to be incapable of working within the totality of a particular technical system. For instance, an internal combustion engine which itself is capable of functioning is not so if used as the underwater engine of a submarine.

Law 1 can be explained in this way: a technical system is capable of life if all its parts do not have a "bad mark", and at the same time the "overall mark" is given according to the quality of work of a given part in the overall system. If at least one part is given a "bad mark" the system is incapable of life even if all the other parts are given "top marks." A similar law applied to biological systems was formulated by Liebich in the mid-nineteenth century ("the law of the minimum".

From Law 1 stems a consequence which is very important in practice.

For a technical system to be controllable it is necessary for at least one of its parts to be controllable.

"To be controllable" means changing its properties in a way required by the controller.

Knowledge of this corrollary allows one to understand the essence of many problems better and to assess more correctly the solutions which are obtained. Let us take, for instance Problem 27 (the sealing of ampules). Given is a system consisting of two non-controllable parts: the ampules are uncontrollable in general - it is impossible (inconvenient) to change their characteristics, and according to the specifications of the problem the burners are badly susceptible to control. Clearly solution of the problem will consist of introducing into the system one further part (S-Field analysis at once suggests that this should be a substance and not a field as in Problem 34, for instance, on painting cylinders). What substance (gas, liquid, solid body will prevent the flame reaching parts it shouldn't and at the same time not interfere with the ampule apparatus? Gas and a solid body can be ruled out, and there remains only a liquid - water. Let us arrange the ampules in water in such a way that only the tips of the capillaries appear above the surface (Pat. Cert. No 264 619). The system acquires controllability: one can change the level of the water - which ensures an alteration in the boundary between hot and cold zones. One can change the temperature of the water - this ensures stability of the system during the working process.

2. The law of "energy conductivity" of a system

A necessary condition for the life capability in principle of a technical system is the unhindered passage of energy through all parts of the system.

Any technical system is a transformer of energy. Hence comes the evident need for the transfer of energy from the engine via the transmission to the working organ.

The transfer of energy from one part of the system to another can be substance (for instance, a shaft, gears, levers, etc), field (a magnetic field, for instance) and substance-field (the transfer of energy as a stream of charged particles, for instance). Many inventive tasks boil down to selecting out this or that form of tranfer which is more effective in a given set of circumstances.

Such is Problem 53 on heating a substance inside a rotating centrifuge. The essence of the problem is in creating an "energy bridge". Such bridges can be homogeneous and non-homogeneous. If a form of energy changes upon transmission from one part of the system to another this is a non-homogeneous bridge. In inventive problems one has to deal most of all with bridges of this kind. Thus in Problem 53 on heating a substance in a centrifuge it is an advantage to have electromagnetic energy (the transmission of which does not hinder the rotation of the centrifuge), but inside the centrifuge heat energy is needed. Specially significant are those effects and phenomena which permit one to control energy at the outlet of one part of a system or at the entrance to another of its parts. In Problem 53 heating can be guaranteed if the centrifuge is in a magnetic field, and inside the centrifuge is placed, for instance a ferromagnetic disk. However, the conditions of the problem require not only heating up the substance inside the centrifuge, but also maintaining a constant temperature of around 250o. However one may change the selection of energy, the temperature of the disk should remain constant. This is ensured by setting up a "surplus" field, from which the disk extracts energy sufficient for heating up to 250o after which the substance of the disk "switches itself off" (the switch through the Curie point). On lowering of the temperature the disk "switches itself on" again.

The corollary of Law 2 is also significant:
In order for part of the technical system to be

controllable it is necessary to ensure conductivity of energy between this part and the controlling organs.

In problems of measuring and discovering one can speak of the conductivity of information, but it often amounts to energy transfer, albeit weak. An example is served by the solution of Problem 8 on measuring the diameter of the polishing disk working inside a cylinder. The solution of the problem is eased if it is regarded not as information but as energy conductivity. Then for solving the problem one needs to answer first of all two questions: in what form is it simplest to conduct energy to the circle and in what form is it simplest to extract the energy through the walls of the circle (or along the shaft)? The answer is quite apparent: in the form of electrical current. This is not yet the final solution but a step has been taken toward the correct answer.

3. The law of harmonizing the rhythms of parts of the system

An essential condition for the living viability in principle of a technical system is the harmonisation of the rhythms (frequencies of vibration, periodicity) of all parts of the system.

Examples of this law were presented in Chapter 1.

To "cinematics" belong laws governing the development of technical systems independently of the concrete technical and physical factors conditioning this development.

4. The law of increasing the degree of idealness of the system

The development of all systems proceeds in the direction of increasing the degree of idealness.

The ideal technical system is one whose weight, volume and area strive toward zero although its ability to carry on functioning at the same time is not diminished. In other words, the ideal system is when there is no system but its functions are preserved and carried out.

Despite the obviousness of the concept "ideal technical system" a certain paradox exists; real systems become ever greater in dimensions and weight. The size and weight of planes, tankers, cars increase. This paradox is explained by the fact that the reserves released in improved systems are directed at increasing its size and, most importantly, at raising the working parameters. The first automobiles reached a speed of 15 to 20kph. If this speed had not increased cars would gradually have appeared which were far lighter and more compact with the same degree of reliability and comfort. However, each improvement in the car (using more reliable materials, raising the coefficient of efficient action of the engine, etc) was directed toward increasing the speed of the vehicle and what "supports" this speed (a powerful braking system, rugged axles, reinforced amortization). In order to see at a glance the growth in the degree of idealness of a car, one needs to compare the modern car with an old record-breaking car which could go at the same speed (over the same distance).

A visible secondary effect (growth in speed, horse power, tonnage, etc) masks the primary process of increasing the degree of idealness of the technical system. But upon solution of inventive problems it is necessary to orientate precisely toward increasing the degree of idealness - is a reliable criterion for correcting the presentation and evaluation of the answer obtained.

5. The law of uneven development of parts of a system

The development of parts of a system proceeds unevenly; the more complicated the system, the more uneven the development of its parts.

Uneven development of the parts of a system is a reason for the occurrence of technical and physical contradictions and hence inventive problems. For instance, when the rapid growth in the tonnage of cargo ships began the power of engines was rapidly increased, but the means of braking remained unchanged. As a result the problem arose of how to brake, say, a tanker with a water displacement of 200,000 tons. This problem has to this day found no effective solution. From the starting braking till coming to a full stop large ships may travel for several miles...

6. The law of the transition to a super-system

Having exhausted the possibilities of development a system is included in a supersystem as one of its parts; in so doing further development takes place at supersystem level.

We have already spoken of this rule.

Let us move on to the "dynamics." This includes rules reflecting the development of contemporary technical systems under the effect of specific technical and physical factors. The laws of "statics" and "cinematics" are universal, they are just at all times and apply not only to technical systems but also to any systems in general (biological, etc). "Dynamics" reflects the main tendencies in the development of technical systems precisely in our time.

7. The law of the transition from macro to micro level.

The development of working organs proceeds at first on a macro and then on a micro level.

In the majority of modern technical systems the working organs are the "irons", for instance, the propellers of an airplane, the wheels of a car, the cutters of a miller's lathe, the blade of an excavator, etc. Development of such organs is possible within the limits of a macro level; the "irons" remain "irons" but become more and more refined. However, the moment comes inevitably when further development on a macro level becomes impossible. The system preserving its function is as a matter of principle restructured: its working organ begins to act on a micro level. Instead of "irons" the work is carried out by molecules, atoms, ions, electrons, etc.

The transition from macro to micro level is one of the main (if not the main) tendency of the development of modern technical systems. Therefore in studying the solution of inventive problems special attention should be paid to examining the "macro - micro transition" and the physical effects which have brought this transition about.

8. The law of increasing the S-Field involvement.

The development of technical systems proceeds in the direction of increasing the S-Field involvement.

The meaning of this law is that non S-Field systems strive to become S-Field, and in S-Field systems the development moves in the direction of a transition from mechanical to electro-magnetic fields; increases in the degree of dispersion of substances, the number of links between the elements and the responsiveness of the system.

Numerous examples illustrating this law have already been encountered in the solution of problems.

STANDARDS FOR THE SOLUTION OF INVENTIVE PROBLEMS

In the last chapter we began to construct a multi-level pyramid of methods: simple methods, paired methods, complexes of methods. The structure gets more complex, the strength of methods increases, specialisation begins to show itself, and "adherence" to this or that class of problems. At the fourth level there should be even more complex methods, distinguishable for their special power and clear specialisation. Such methods have been discovered, they constitute the store of standards for the solution of inventive problems.

A few words about the name. It has a certain demonstrative value. Of course, one could replace the word "standards" by "solution of typical problems" or "certain characteristic classes of problems and their typical solutions". But the word "standard" expresses the basic idea in a shorter and more accurate form, that there are complex ideas which have to be applied by dint of force, because for their classes they guarantee solution on a high level.

And so, the fundamental features of standards consist in that:

-they include not only methods but also physical effects;

-the methods and effects included in the standard form a definite system (i.e. are linked not just "as it comes" but in a definite order);

-the system of methods and effects is distinctly directed at eliminating physical contradictions, typical for the given class of problems;

-clearly visible is the link between standards and basic laws of the development of technical systems.

Breadth, common identity between solution and effectiveness - these are absolutely necessary requisites for any "candidate" standard. Let us take, for instance, the application of the Toms effect. Use of this physical effect always leads to solutions of a high level. But the field of its action is very narrow: in essence it amounts to a single problem, that of how to reduce the friction of a liquid and a solid body while they are in a state of relative movement. Pat. Cert. No 412 382 suggests an additive of long-chained polymers to a liquid "for jet action on solid materials"; in Pat. Cert. No 424 468 the same effect is patented as a "means of working a liquid-ring machine, such as a compressor"; in Pat. Cert. No 427 982 long-chained polymers are introduced into lubricants for reaming pipes; Pat. Cert. No 464 042 is the same thing, but here it concerns an "electrical water-filled machine". There is a multitude of such machines, but everywhere the task is the same, that of reducing the friction of a liquid against a solid surface. Inventive solutions based on the use of one physical effect rapidly become trivial. The use of the electrohydraulic effect at the end of the 1940s produced strong inventive solutions, but in ten years this method had become trivial. Standards do not refer to a concrete physical effect, but to the type of effect, and for this reason standards have a considerably extended life. Certain of them call on physical effects, which will be discovered in the future.

The description of one standard includes a detailed justification for it and numerous examples, reflecting the nuances of application. We shall briefly examine only the essence of the first ten standards (around 50 standards have been worked out already but many of them have not yet completed their "experimental period").

Standard 1. If it is difficult to discover an object at any given point in time and if one can introduce additives to it in advance the problem can be solved by preliminary introduction into the object of additives which make an easily exposable (most frequently an electromagnetic) field or make it easily interact with an outside environment, revealing itself and hence the object. By analogy measurement problems are solved if they can be presented in the form of the consequences of problems for exposure.

Examples of this are the solutions to Problem 9 (the addition of luminescent traces to paints) and 10 (the addition of ferromagnetic substances to polymers). According to Pat. Cert. No 415 516 the temperature in places difficult of access is measured by introducing a diamond chip: with the change in temperature the reading from broken light passing through the diamond is changed. The essence (in the S-Field sense) in all these cases is the same: one is given one substance, a second is introduced which is "able" to interact very well with an outside electromagnetic field.

Standard 2. If one needs to compare an object with a template in order to discover deviations the problem is solved by optical matching of the image of the object with the template or with an image of the template, whereby the image of the object should be the opposite in colour of the standard or its image. Problems of measurement are solved in a similar way if one has a standard or an image of one.
An example is provided by the solution to Problem 28. Another example is Pat. Cert. No 350 219; a disk with perforated openings is controlled by comparing a yellow image of the disk with a blue image of the standard. If on the screen a yellow light appears this means that a perforation is missing from the control disk. If a blue light shows up it means that the disk has a perforation too many.

Standard 3. If two substances moving relative to each other have to touch and in so doing a harmful effect arises, the problem is solved by introducing between them a third substance which is a variant of one of the substances given in the specifications of the problem.

An example of this is the solution to Problem 14. Standard 3 can also prove a solution to Problem 17; at the place of a bend in a pipeline a permanent magnet is set up outside; a layer of balls is "glued" to the inside wall; the moving balls collide not with the walls but with the fixed balls; if one fixed ball is knocked out of position its place is taken by another (Pat. Cert. No 304 356). Problem 36 is solved in a similar way. In Problem 21 a harmful phenomenon (erosion) occurs between water and metal.: one introduces "altered water", i.e. a layer of ice: parts of the underwater pod which must be protected are cooled and on them grows up a fine and constantly renewed layer of ice (Pat. Cert. No 412 062).

Standard 4. If it is necessary to control the movement of an object one should introduce into it a ferrmagnetic substance and use a magnetic field. Problems of ensuring the deformation of substances should be solved in a similar way, for treating its surface, splitting up, mixing, changing viscosity, porousness, etc.

Examples are found in the solution to Problem 6 on changing the properties of soil on a testing ground and Problem 28 on transferring the lines of a drawing. In Pat. Cert. No 147 225 ferromagnetic particles are introduced into inks and controls these inks with the aid of a magnetic field. According to Pat. Cert. No 261 371 a ferromagnetic powder is introduced into a catalyst and controls its movement with the aid of a magnetic field.

Pat. Cert. No 4333 829 describes a muffler using a ferromagnetic liquid which hardens in a magnetic field; in Pat. Cert. No 469 059 the change in viscosity of such a liquid in a magnetic field is used for controlling silencer equipment. There are very many similar inventions all relating to solutions of a high level.

Standard 5. If it is necessary to increase the technical specifications of a system (mass, dimensions, speed, etc) and this comes up against obstacles of fundamental importance (a ban imposed by natural laws, the absence of the necessary substances, materials, power in the present state of technology, etc), the system must be introduced as a subsystem into another more complex system. The development of the original system ceases and its place is taken by a more intensive development of the more complex system. An example of this is served by the creation of a gas-heat protective suit.

Standard 6. If it is difficult to perform an operation with delicate, fragile and easily deformable objects, for the duration of the operation the object should be joined to a substance which makes it firm and stable; later this substance can be removed by dissolving, evaporating, etc. According to Pat. Cert. No 182 661 thin-walled tubes of nichrome are made (by paring) on an aluminium rod and the rod is removed with an alkali.

Standard 7. If one needs to combine two mutually exclusive actions (or two mutually exclusive states of an object) each of them can be made discontinuous and linked in such a way that one action is performed in the pauses left by the other. In so doing the switchover from one action (state) to another should be implemented by the object itself, for instance, by using phased switchovers emanating during changes in external conditions.

Examples - the solution to Problem 25 on the lightning rod and Problem 48 on protection against icing up.

Standard 8. If it is impossible to measure directly the state (mass, dimensions, etc) of a mechanical system, the problem is solved by triggering off in the system a resonance vibration and from the changes in oscillation one can determine the changes which are taking place.

The frequency of inherent oscillations is the pulse of the technical system (or part). The ideal method of measurement is when no sensors are employed and the system itself reports back on its condition. According to Pat. Cert. No 244 690 from the inherent frequency of oscillation the weight of a moving thread can be determined (previously it had been necessary to cut off a part of the thread and weigh that).

Standard 9. If it is necessary to increase the technical indicators of a system (accuracy, speed, etc) and this clashes with obstacles of a fundamental kind (a ban imposed by the laws of nature, a sharp deterioration in other properties of the system) the problem is solved by switching from a macro to a micro level; the system (or a part of it) is replaced by a substance capable of carrying out the necessary actions in conjunction with a field .

Standard 5 concerned transferring from a system to a supersystem; the essence of Standard 9 is to transfer from a system to a subsystem. The reader is already familiar with examples. In particular, according to Standard 9 Problem 52 (for the creation of a super accurate tap one needs to use heat distribution, magnetic striction, a reverse piezo effect).

Of great significance for the application of standards is the possibility or impossibility of introducing additives, in accordance with the requirements of the standards 1,3,4 and 6. So far we have used the words "the object can be changed","one cannot change the object". Now these words are given a specifically physical meaning, which enables one to use more accurate definitions: "one can introduce additives" and "additives cannot be introduced." The degree of difficulty of a problem in many ways depends on this "can" and "cannot". Therefore Standard 10 relates specially to the translation of "cannot" and "can".

Standard 10. If one needs to introduce additives, and this is forbidden by the terms of the problem one has to take roundabout routes:
1) Instead of a substance one introduces a field; 2) instead of "internal additives" one uses "external"; 3) an additive is introduced in very small doses; 4) an additive is introduced temporarily; 5) as an additive one uses part of a substance already present, transformed into a special condition or already in such a condition; 6) instead of an object one uses a copy (model) of it, into which the introduction of additives is permissible; 7) additives are introduced in the form of chemical combinations from which

they can later be separated out.

The roundabout routes 2 and 4, for example, can be used for solving Problem 20. Onto the diamond is dusted a fine layer of metal and one carries out the orientation of diamond chips with the aid of a magnetic field. On being polished the unwanted dust layer is wiped away immediately.

In solving a problem by the trial and error method, you may unexpectedly come across the solution. One minute there is no answer - the next minute there is. This unexpectedness is reflected in many expressions such as "enlightenment", "dawning", "eureka", "insight". The words are different, but the meaning is the same: the solution appeared suddenly and the fog instantly cleared to light.

In reality in working by the trial and error method the replacement of fog by light takes place in an elusively short space of time. For the psychologist studying the inventor's creativity on the trial and error level, "enlightment" is one of the fundamental phenomena. It is another thing if the psychologist is studying inventive creativity conducted on ASIP level. In this case the absence of "enlightenment" is just as firmly rooted and fog is gradually replaced by light. Here is a tape transcript of the solution of Problem 54 on the lost diamond (the problem was solved by a mathematician who had graduated from a Public Institute of Inventive Creativity): "This is a problem of discovery. Hence, one needs to apply Standard 1: one has to introduce some kind of additive to the diamond. All right, but it is impossible to introduce an additive! Here is the contradiction. In this instance we can turn to Standard

10 - introducing an additive temporarily or in very small
doses. But this doesn't work either, there is no way of
introducing an additive. Next comes the roundabout route -
use as an additive something which is already to be found
in the substance. But what is there? Diamond is a crystal,
a crystalline network. Here we would have to break down
the crystal network? We have to! That is to say, they have
to be used as a mark. Like birthmarks in a person. Nothing
can be done with the stolen diamond, but for all others
one should make an X-ray picture in advance. What would
emerge would be something like dactylloscopy for
diamonds..."
In solving a difficult problem by eliminating variants,
the inventor may not move from the spot for years; which
may mean that of 50,000 variants he has already looked at
3000. Solving it in the ASIP way is another matter
entirely. A person consciously guides the process of
solution, calling upon the knowledge of this or that
particular law, method, approach, etc. Each operation
brings solution near and clears the fog. The outlines of
the solution take shape gradually,; (and of course far
more quickly than with the trial and error method).
According to tradition "enlightenment" is customarily
considered to be an inevitable feature of creativity: if
there is "enlightenment" there is creativity; no
enlightenment - no creativity. Now, at the new level of
the organization of creativity instead of "enlightenment"
or "dawning" a psychological attribute of creativity has
become "clearing" - the gradual changeover toward the
light.
In this process (there is a kind of paradox here) the

solution of problems is partially known even before the problem is presented. Without knowing the problem we know in advance the laws, i.e., the answer in general form. The process of solution consists in the transition from general laws to their specific elucidation in a given case.

Standards for the solution of inventive problems can be applied before analysis (at step 1.7). But it is more effective to use them after analysis, in any case after building the model of the problem, therefore standards enter the table of typical model problems and S-field transformations.

Sometimes for solution of tasks it is necessary to consistently utilise several standards.

PROBLEM 56

An apparatus for packing fruit in cardboard boxes includes a vibrating table onto which the boxes are placed (the vibration enables one to increase considerably the compactness of packing. The fruit enter from above along a chute. Unfortunately tender fruits are banged on falling (let us say that they fall 0.5 meters). Lowering the chute to the bottom of the boxes and then lifting it with some kind of device is too complicated and therefore bad. what can be done?

One peach hits another and this has a damaging effect - this is a typical problem for Standard 3. One needs to introduce between the two colliding peaches a "soft peach", i.e. some kind of elastic balls made, for instance, of polystyrene (such balls, thanks to the vibration would be at the upper layer of fruit). After filling the boxes it would be necessary to remove the balls - and this is a case for Standard 4. Into the balls are placed ferromagnetic disks; after the boxes are filled an electromagnet placed above the boxes is activated, the balls "leap" out of the box, an empty box comes along, the magnet is switched off, the balls are released. The effectiveness of the solution (Pat. Cert. No 552 245) is achieved by the joint application of two standards; one uses balls as shock absorbers and ensured a method of controlling them.

THE WOOD FOR THE TREES

The strongest means of solving problems (S-Field analysis, standards) are simultaneously instruments for elucidating new problems. The forecasting function is also inherent to methods used at step 6.3. Let us examine this using a concrete example.

Suppose that for the first time an electromagnetic metre has been made (an consumption meter) of the flow of electro conductive liquid. The principle of constructing such an consumption metre is extremely simple (fig. 16,a): into the stream to be measured (P) are introduced electrodes (E), a magnetic system (M) is placed outside setting up a magnetic field; the flow intersects the magnetic power lines, and at the electrodes an electromotive force arises. If now one asks the question:

"Suggest new constructions of electromagnetic consumption metres", the search using the trial and error method would not bring rapid results, because it is not known how to change the existing pattern. Let us use the simplest method : re-arranging the parts. The structure of the original construction can be denoted as: the MESEM. At the centre is the stream, along both sides of the stream are electrodes, on the outside the magnetic system. It is obvious that by rearranging the parts one can obtain another five constructions: EMSME (fig 16,b); SMEMS (fig. 16,c);MSESM (fig. 16,d); ESMSE (fig. 16,e); SEMES (fig. 16,f).

By the moment that such a morphological analysis had been made for the first time, only the chute consumption metre along the MESEM pattern together with the log (speed gauge) with the SEMES pattern were known. Four patterns turned out to be new, which had their own characteristics and advantages. For instance, the MSESM pattern allowed one to measure the local comsumption along the breadth of the stream. The log with pattern SMEMS works on the internal magnetic field of a solenoid and for this reason more sensitive that the SEMES log, working according to the dispersal field.

In this way even the simplest methods (the rearrangement of parts) can be used not only for solution of problems, but also for elucidating the field of application of the principle arrived at, i.e. for the purposes of forecasting.

Fig. 16

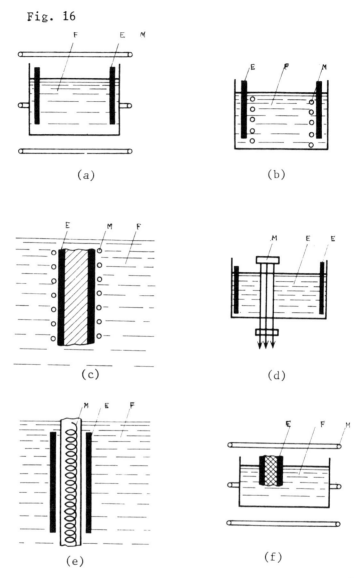

(a)

(b)

(c)

(d)

(e)

(f)

Let us examine, for example, the magnetic filter (Problem 13). It includes a magnetic system (M), a ferromagnetic powder (F), through which the stream of dusty air (let us denote this stream with the letter A - an artefact). The structure of the filter is MFAFM.

It is clear that another five structures are possible: FMAMF; AMFMA; MAFAM; AFMFA; FAMAF.

For instance, the "Electromagnetic filter; in order to lower the mean consumption of energy and increase productivity a filtration element made of grainy magnetic material is placed around the source of the magnetic field and forms an outward closed magnetic profile." (Pat. Cert. No 319 325). This invention (an internal magnet) appeared only seven years after the ordinary filter (the outward magnet) had been invented)...

Let us consider six possible structures of the length of the table and on the side let us note five possible states of artefact: gas, liquid, solid body (a steel rod, for instance, and elastic body (rubber) a powder. A table is obtained contained 30 cells, and they contain not only the pattern of the filters but also the patterns of technical systems which are different in function. For instance in the cell at the intersection of the MFAFM column and the "solid body" line one could place the invention in Pat. Cert. No 499 912 on "A method of non-filtrated drawing of a steel wire, including deformation by stretching; in order to obtain a wire with a constant diameter the necessary deformation is brought about by drawing the wire out through a ferromagnetic mass, placed in a magnetic field." In the CSSR (Czech) Patent No. 105 766 a magnetic bung is described installed in the sump of an engine to extract metal particles from the oil: the structure is IFMFI, and the aggregate state of the artefact (oil) is liquid.

In the table only two methods are employed: the

rearrangement of parts and the change in the aggregate state of the artefact. One can use a third method: the transition from a progressive movement (of the artefact, powder or field) to a rotating one or build two 30-cell tables for two sorts of movement. In this way the originating pattern according to plan recycles in 60 patterns.

With a detailed treatment the number of patterns can be considerably increased. For this it is necessary to consider which part of the system is moving and which is immobile. "In 1962 Gennadiy Shulev, at that time a post-graduate student at the Kaliningrad Institute for Fisheries and the Fishing Industry, presented a new magnetic-abrasive technique for processing metals. For this in May of the same year he was granted Pat. Cert. No 165 651. He had cleverly modernised the idea of processing metal which had been put forward in our country in 1938. The idea had been put forward but without results, for at first it had been proposed to process cylindrical surfaces in a rotating magnetic field. Shulev developed the idea that the magnetic field should be stationary, and the part should rotate. It was tried out on a lathe and worked (Tekhnika i nauka, 1976, No 7, p.15). Twenty four years for the transition to the idea of "don't turn the field, turn the artefact" is the price paid for unorganised thinking. According to ASIP-77 the development and transformation of ideas can be obtained by the introduction of the third axis of the "Use of Physical Effects and Phenomena"; the control of magnetic characteristics by means of changing the temperature of a system, the transition through the Curie point, the

Hopkins and Barkhausen effects. For instance, in Pat. Cert. No 397 289 on contact welding of a ferromagnetic powder to the working surfaces of parts (for ensuring the even distribution of the delivery of the powder) it is heated to the Curie point. Two three dimensional tables - there are already more 200 of these. Behind the wood of ideas one can clearly see a whole grove of ideas...

The construction and filling of similar tables is a fine exercise for developing "ASIP-style" thinking, and since 1976 has been used systematically at study sessions in public schools and institutions of inventive creativity. Not infrequently these sessions turn up interesting inventions and new directions for research and development are revealed.

PROBLEMS

You can easily solve Problems 57 to 59 although they are extremely difficult for anyone who has not yet heard of standards. But with Problem 60 one has to work at an unhurried and patient rate. The solution of this problem can be turned into an interesting piece of research.

PROBLEM 57
Pat. Cert. No 206 207 describes a lathe for tapping threads; in order to take undue strain from the tool the lathe is equipped with a liquid-filled chamber and floats connected to the spindle block. According to Pat. Cert. No 354 297 a similar floating arrangement is used

for taking the weight off bearings in equipment for measuring rotating momentum. The bigger the floats the greater the hydrostatic force developed. But increasing the dimensions of the floatation chamber is inconvenient. Using a heavy liquid instead of water is expensive, inconvenient and dangerous. Make a forecast of the direction which such float chambers might take.

NB. If difficulties arise go back to Problem 50.

PROBLEM 58

There is an apparatus for obtaining a polymer in the shape of miniscule spheres. The apparatus is a tank containing molten polymers. Leading to its surface is a pipe carrying compressed air which picks up and pulverizes the polymers. The air flow with the polymer droplets is carried along the pipe: the droplets congeal, fall to the lower wall of the pipe and roll into a special container. Unfortunately the apparatus produces too many large-size spheres. Various different ways of introducing air have been attempted but have not brought an improvement. A latticework has been built into the pipe, nets - but this resulted in a fall in productivity.

What standard should be applied here? What is the solution of the problem following this standard?

PROBLEM 59

Conserves are filled into litre jars with
metal lids. It is necessary to checkl whether
the lid fits tightly enough to seal the opening
of the jar. To do this the jars are lowered into
a water-filled vat and watched to see whether
air bubbles appear (the water would enter
through any leaks and expel air). This method is
slow and unreliable.

What standard should be called upon? What
solution does this standard produce?

NB. The problem is one of discovering drops
of water entering the jars. The model of the
problem puts it in Class 1 where one substance
is given. The contents of the jar, the jars
themselves, the lids are not included in the
model of the problem.

PROBLEM 60

Convert the answer to Problem 59 into a
table showing the "Disposition of Parts - the
Overall States of the Artifice." Use the
examples of inventions quoted in the text of the
book to fill in certain boxes in the table. Try
to fill in empty boxes with patterns. What
application can these patterns have? Are there
no new inventions among them?

8

The Science of Invention

"WANDERING WITH A DISTRACTED AIR..."

Can one, after all this, invent without a theory? Without S-Field analysis, without meticulous ASIP operations, without all this difficult science? You can. Here is a typical example.

Steel is poured into converters, huge metal crucibles with an inside lining (facing) made of heat-resistant brick. Every 7 to 10 days the facing burns out, works has to stop, the converter has to cool off, the lining has to be broken out and renewed. The idea arose of changing the crucible of the converter altogether. But the converter is as a high as an eight-storey house; it is too complicated to wheel out the crucible on rollers and then put a new one in. What can be done? The inventor V. Gorelov, a Candidate of Technological Sciences, pondered long over this problem. The problem could not be solved and many "void" trials were carried out. Until one day... "I couldn't avoid buying a toy for my daughter," Gorelev relates, "and so wandering with a distracted air through the store I fancied some Russian nesting dolls, lined up prettily behind their "mama," into which they would all be able to fit. They reminded me of something, but what? I went past, and then came back. I had remembered! If you were to remove the top part and turn it upside down this is a converter into which a slightly smaller body could be fitted with a lining" (Izobretatel' i ratsionalizator, 1976, No 9, p 20).

Well, but what would have happened if V. Gorelev had not gone into the toy shop? Or if on that day no nesting dolls had been on display? The problem would have been solved one way or another, of course, but after a lot of time had been wasted."

Let us turn now to the table of fundamental methods [13]. One needs to increase the "convenience of repair" (line 34). If one takes the known paths (by removing the entire crucible), more time is needed for installing the repaired crucible. Column 25 ("time spent") or 32 ("convenience of manufacture"); the methods in the first box: 32,1,19,25; in the second: 1,35,11,10. Method 1 - is to divide the object into parts. This is precisely the method hit upon by V. Gorelev, while "wandering with a distracted air2. (The nesting doll had already been implemented in the construction of the crucible, since the lining was inside the body. The essence of the invention consisted in removing the lining, therefore "the distracted air" should have been fixed on the dismantled doll).

Another task and another inventor: Candidate of Technological Sciences A. Belotserkovsky. "... for the ideal liquid for hydraulic extrusion two mutually exclusive demands have to be met: in its zone of action for preparation the liquid should be non-viscous and should transmit hydrostatic pressure well, but in the zone of sealing and friction (where the plunger enters the container) the liquid should be highly viscous with good lubricating properties. We made numerous attempts to combine such a liquid of various components, we turned to the chemical institutes, patiently studied the literature and patents but did not succeed in finding a suitable liquid. The solution came unexpectedly and at an institution most unsuited for scientific creativity, a cocktail bar. One Saturday evening we were distractedly looking at the manipulations of the lady bartender who was skilfully pouring multilayered drinks. At that time a stupidly simple idea came to me: what if a "cocktail" were to be made in the container for hydraulic extrusion too (Pat. Cert. No 249 906). We tried it out and really it all went off excellently" (Izobretatel' i ratsionalizator, 1970, No. 12, p. 21).

The physical contradiction is apparent, and the solution of the problem is optimally eased. It is clear at once that the contradictory properties should be separated in space. Consequently there can be two and only two possibilities: 1) to take a single liquid and create different conditions on different parts of the machine; 2) take two liquids and separate them in space in such a way that the two liquids do not intermingle. The inventor could for a long time see on the first path, and that only partially: he sought a liquid which in different parts of the machine would behave differently. Any student at a public school of inventive creativity knows that there are liquids which change their viscosity in a magnetic or electrical field. There are liquids which change their viscosity depending on the temperature, on the gradient of speed in layers... But it is simpler to choose the second route: have different liquids work in different parts of the machine. For this it is necessary to select non-mingling liquids (a "cocktail") which are viscous and non-viscous. It is interesting to contrast the account of the inventor with the long since published commentary to a study problem [12, p. 176]:

"We poured into the container for hydroextrusion ... a non-viscous operating liquid, to put it simply, water... Into the bulk of the container we poured a layer of highly viscous liquid, whose density was lower than that of water... this was SU type mineral oil".

"The main stipulation is that it should not mix with oil... All substances are either organic or inorganic. It was already known to the alchemists that "like dissolves in like". Oil is an organic substance. This means that the screen should be inorganic. The most widespread inorganic liquid is water."

The final answers coincide: water and vegetable oil, water and petroleum oil. In both instances the composition of the "cocktail" is the same.

At times it is difficult to understand why in all searches ideas "wander" in the first instance - in books on the theory of the solution of inventive problems. After all, one can simply pick out 40 basic methods and you do not even have to learn this...

Engineer Yu. Portnyagin came up against a problem very similar to the one solved by V. Gorelov: it was necessary to speed up the repair of a furnace, on this occasion for glass blowing. The bottom of the furnace was clad in fire-resistant fire-clay bars - which took up 90 percent of the time spent on repairs. The problem was that the bars had to lie close to each other, and therefore they had to be turned by hand. Yu. Portnyagin had attended some TSIP lectures, and knew that there are 40 methods. Therefore the solution of the problem took a different course. "Having selected the methods of solving inventive problems," Portyagin writes, (Izobretatel' i ratsionalizator, 1973, No. 4, p. 28), I was constantly racking my brains over how to reduce the gaps between the bars to a minimum. One method was to strengthen a harmful factor and turn it into a useful one... I mentally increased them, making them wider and wider... What next? This is what: one had to smother all this space with its uneven curved walls with highly fire-resistant concrete... Our economists calculated the economic effect of making the job easier as 103 roubles 33 kopeks for each cubic meters of facing! The Committee for Inventive Matters awarded me for this method an author's patent No 270 209.

It is simple enough to look at the list of methods but how seldom it is done! The chief designer of the project, D. Tsitron tells about this invention, based on the application of the "air cushion"; as we have seen this is an extremely widespread method, and in the literature on TSIP it is not difficult to find examples of its use. But D.Tsitron discovered it not in books but ... among his son's toys: "One day while my son and I were sorting through his old toys I saw an inflatable rubber clown. That was just what I was looking for!" (Izobretatel' i ratsionalizator, 1972, No. 5, p. 12).

And here is another curious example. Let us present it in the form of a problem:

PROBLEM 61
There exist so-called metal-plating

lubricants. They are 90 percent ordinary oil and 10 percent finely ground metal powder. In operation this powder applies to the abutting surfaces the finest protective layer of metal. The making of the lubricant is simplicity itself - mechanical mixing.

But such lubricants are unsuitable if the gap between the abutting surfaces is smaller than the granules of powder. One could, of course, grind the powder more finely and make a colloidal solution. But even in this case one would obtain particles that were too big. Grinding more would mean switching from a colloidal solution to a true solution (in a true solution) the metal would be held in the form of molecules, atoms or ions). But metals are not soluble in oil, so what can be done?

There is a clear physical contradiction: metal has to be present in the lubrict in order that plating take place, and the metal should not be in the oil because it should not contain large particles. This is the most typical problem involving the application of Standard 10. This standard includes seven methods of introducing a substance, when it cannot be introduced. According to the specifications of the problem six methods are ruled out at once (the introduction of a field instead of a substance, the introduction of an "external" additive, etc. Only one remains, that of introducing an additive in the form of a chemical compound from which it can be later extracted.

What kind of compound can this be? Evidently it

should possess three properties: it should contain metal, dissolve in oil (to produce a true solution) and shed its metal when frictional forces are exerted.

It is easy to meet the first requirement. There are as many substances containing metals as you would wish, common salt, for instance. It contains metal - and it does not (it is bound and does not display its metallic properties). The third requirement is also easy to satisfy: substances containing metal decompose, for instance, under the effect of an electrical current or heat field. The latter is better, since current has to be conducted whereas heat is produced itself by friction. Now it has to be determined which substances dissolve well in the lubricant. But even here there is no difficulty. The lubricant is an organic substance, and organic substances dissolve in it well ("like dissolves in like"). This means that one needs organic-metal substances (a certain metal compound and an organic acid) which dissolves at the temperature of friction we require. It remains for us to take the handbook of organic chemistry and select a suitable organic-metal compound.

This task was solved on many occasions at TSIP study sessions. Obtaining the general idea for the solution ("One needs such and such an organic-metal compound) takes up on average 30 to 40 minutes; around 40 percent of answers given in the first year were correct. In the second year it was around 70 percent. Selecting a specific substance is usually given in the form of homework. On average around 60 percent handed in homework responses (chemistry handbooks were essential and not everyone managed to get hold of them), and practically all of these contained correct answers (sometimes students indicated not only the substance required, but two or three others as well).

Now let us look at how this problem was solved "in the field". An extract from an article by the inventor, Candidate of Tech. Sciences V. Shimansky: "To be brief, one needs an oil-soluble deposit... The most varied compounds were investigated. Too many to name. Once again chance helped out. Once I was standing in a book store and heard someone say to the sales girl: "Please give me 'Methods of Elementary Organic Chemistry'. A happy idea! And what if one were to try out organic-metal compounds? I asked for the 'Methods' for myself. This was a veritable gold mine. Acetic acid cadmium, it appeared dissolves at the 250o. An experiment produced my answer: this substance dissolves in oil. The friction surfaces are coated in cadmium."(Izobretatel' i ratsionalizator, 1972, No. 2, p. 6).

Perhaps the most tragic thing about these stories is that the lesson for the future has not been learned from them. The very same inventors in solving other problems operate in the same old trial and error method once more wasting whole years through ignorance of the elements of the TSIP. "Three years were spent," writes Candidate of Tech. Sciences A. Belotserkovsky about another problem, "in which I tried out various ways of changing the form of a process, but without success..." (Izobretatel' i ratsionalizator, 1972, No. 8, p. 6). If the team of engineers headed by a Candidate of Tech. Sciences, spent three years on "various ways of changing the form of a process" and "wandering with a distracted air" this must have cost the State no less than 50,000 rubles. By way of comparison: the annual budget of a public institute of inventive creativity (with 100 students) comes to only 3000 rubles.

We have spoken of inventors tackling problems alone or in small groups. Perhaps things are different in very large teams? Perhaps they have a more efficient technology of creativity?

This is what General Designer O.K. Antonov has to say (Literaturnaya gazeta, 14 August, 1968, p. 2):

"When the 'Antey' was being designed especially difficult was the tail unit design. A simple high fin with a horizontal tail above it despite all the clarity and deceptive attraction of this design recommended by aerodynamicists was impossible. The high vertical fin would wrench the skin of the fuselage of a plane with a huge cutout for the cargo hatch four metres wide and seventeen metres long.

Dividing the horizontal tail unit and suspending "disks" to the ends of the stabilizer was also impossible since this sharply reduced the critical speed of tailplane flutter.

Time went by but the design of the tail unit was not found."

The modern aviation design office consists of a team working according to plan on an overall programme. The general designer does not ponder over a problem in isolation. Every point of the plane is studied by a group of talented designers commanding the most up to date information on everything relating to their special field. But if one such group comes to a halt it throws the rhythm of the whole team out. It is not difficult to imagine what the simple phrase "Time went by but the design of the tail unit was not found" cost. "But once," continues O.K. Antonov, "waking up one night, I started as was my habit to think over the main thing that was troubling me. If the half-disks of the tail unit located on the horizontal tailpiece caused flutter because of their mass they should be placed in such a way as to make this negative factor a positive feature. This means, we had to move them radically and place them in front of the rigid axes of the horizontal unit...

How simple!At that point I stretched out my hand to the night table feeling for a pencil and notebook and in great haste I rushed down the design I had found. In a sense of tremendous relief I fell into a sound sleep on the spot."

Three years before O.K. Antonov's article was published the Ekonomicheskaya gazeta had devoted a sixteen page supplement to materials under the title "Pay Attention: Algorithm of Invention!" [25]. These materials contained in particular a small table on the elimination of technical contradictions. One needs to increase the area of the tail unit - this is the third line of the table; if you take certain routes a harmful factor arises: this is in column 14 of the table. At the intersection in the appropriate box of the table three methods are indicated the first of which word for word coincides with that found by O.K. Antonov: "Harmful factors can be utilized to obtain a positive effect" (remember the problem solved by Yu. Portnyagin). Such tables had been published even earlier and it was by no means necessary to work through a multitude of variants, wasting time and looking for solutions in sleepless nights.

"Wandering with a distracted air" as a method of solving inventive problems not only leads to a waste of time but inevitably throws up a multitude of weak, unsuccessful inventions which are impossible of implementation in practice. Here is a typical example: In our country each month around 300 million items of porcelain are produced. After the first firing the wares are separated into three groups, each of which is then fired for a second time according to their technology. Sorting is carried out by sound: the girls take a plate, tap it with a tiny metal hammer and depending on the tone place it into one of three piles. This kind of sorting is extraordinarily monotonous and difficult. Naturally the inventive task arose of how to get rid of manual labour. And so a group of inventors works out a "robot with arms". One arm of the automaton grasps the plate, the other taps it with a hammer: the sound vibrations are picked up by a microphone and analysed: in short this is a blow by blow copy of the actions of a human being. Books on TSIP contain many examples - a steam-driven oared ship, a legged locomotive, an "armed" sewing machine all illustrating the rule that it is impossible to mechanically copy the actions of a human being. An "armed" sorting machine was built and they tried to put it to work... and discovered a mass of faults. The machine sharply increased the breakage rate; the crude manipulators of the machine were only an outward copy of the human arm which is in fact part of the "arm-brain" system. The machine was not installed in practice; the money spent making it turned out to have been pure waste.

From the TSIP viewpoint, this is an archetypal case of poorly organized creativity. Checking the quality of firing of plates is an unsolved problem. But, in other branches of technology perhaps similar problems have been solved and even perhaps with even more rigid specifications regarding productivity and precision? Such checking is provided for even in the early modifications of ASIP (in ASIP-77 step 1.8,b). If one is to take radio technology alone. Resistors are widely used in radio and they are also made of ceramics, they have to be fired and checked. But resistors are "plates" so small that they cannot be checked with a hammer. Take the ACS-1 (Automobile Compression Station): the ceramic plugs are illuminated by two monochromatic rays of light, and the state of the firing is judged in accordance with the intensity with which the stream of light passes through the sample.

Could it be that somewhere there is a method of inspecting the firing of even finer products? There is indeed. The sun "fires" a seed and therefore in agriculture and the food industry they also have to determine how this "firing" is proceeding. In Pat. Cert. No 431 431 we read: "A method of analysing the structure of a wheat seed by utilising its optical qualities; in order to raise the precision of analysis the translucent and reflective qualities are determined and a judgement is formed on its structure from their correleation."

Once invention is carried out while "wandering with a distracted air" the chances are very few that one's look will turn to this very patent (out of many hundreds of thousands) which would give a hint as to the path of solution. One's look is arrested by what is obvious to everyone... and so we have a machine "with arms".

Difficulties in implementation in practice are inevitable if the invention is significantly ahead of its time. There are difficulties caused by conservatism, an unwillingness to take risks, etc. And nevertheless in many instances the primary cause of difficulties is that the problem has been badly solved.

A FEW COMMENTS ON THE LITERATURE OF INVENTIVE CREATIVITY

In O.K.Antonov's article quoted above it he correctly noted that "much has been written about the process of creativity, including much nonsense." Too true, alas. "Enlightenment","dawning","chance finds" - all these are lightweight expressions which are outwardly appealing. Hence the stream of stories on "wandering with a distracted air". There are five, ten, fifty such stories and the schoolboy, the student, the young engineer is imbued with the conviction that this is exactly the way that one should go about inventing.

Admittedly in recent years definite progress has been made. Books and articles have started to appear on methods of activating the search including the excellent book by G. Jones [26]. As we have noted earlier one should not expect too much from methods of activating the search. They can only improve the ordinary screening of variants. But publication of such methods forces one to turn one's attention to the imperfect state of existing technology - and this is a good thing of itself.

Heightened interest in new methods of solving creative problems at times leads to the appearance of fanciful purely mental methods. Such, for instance, is the "strategy of the seven-sided search" put forward by G. Bush. "It is known," he writes, "that a man should subject up to seven subjects, elements, concepts, and ideas to effective simultaneous examination, comparison and study. In this sense a system based on the 'magic' number seven has considerable advantages over the decimal system." [27,p.90]. There are furthermore seven stages, in each of which there are seven key processes... An inventor, however, need not look at all seven stages at once, since the stages represent the consecutive stages of the process. The "magic" of the figure seven has absolutely nothing to do with it. On this basis one could just as well insist that every city has seven districts, that every district has seven quarters, that every quarter has seven houses, that every house has seven floors, that every floor has seven flats. Behind the "magic" form of the "strategy of seven-sided search" is concealed a hopelessly out of date notion. For instance, on the elucidation and formulation of an inventive problem it is recommended to ask the seven questions posed by the Roman orator Quintillian (1. A.D): who?, what?, where?, etc.

Typical of such "alchemy" (no matter whether it is concealed behind "magic" or super-modern cybernetics terminology) is the total break with study of the objectives laws of development of technical systems, the study of patent information, the solution of specific inventive problems.

Reliable criteria exist for judging a new work on inventive creativity:

1. Is it founded on the study of a sufficiently large mass of patent information?

Serious work cannot be based on a few casual facts. Practice has shown that good research into the theory of inventions is usually built on analysis of no less than 10,000 to 20,000 author's copyrights and patents.

2. Has the researcher taken account of the existence of various levels of inventive problem?

Unless numerous inventions of lower levels are not "sown" comparatively few inventions of higher levels will "take on" and conclusions will be produced valid only for simple problems.

3. Have the recommendations been tried out on a sufficiently large number of new problems at higher levels?

Some time ago R.I.Deryagin [28,p.59] recommended according to the model and form of ASIP an algorithm for the solution of inventive questions (ASIQ). In the work not a single example of the use of the ASIQ for the solution of scientific problems was quoted. Naturally no one paid any attention to ASIQ.

Creativity acquires ever greater social value. This also gives rise to speculative sorties. The old terms "dawning", "enlightenment" and others are sometimes replaced by modern "cybernetic terms. Patterns appear with arrows, circles and the captions "entry", "exit", "information processing block", "accumulator" etc. Behind these, alas, is often concealed, alas, no reality.

Inventive creativity is a complex subject for research. For each "quantum" of knowledge no matter how small, one has to pay with a huge amount of labour. But there is no other way.

MASTERPIECES ACCORDING TO FORMULA

The TSIP teaches us to solve inventive problems "according to formula" and "according to rules." The paradoxical situation arises in which man makes high level inventions (i.e. obtains a high quality product of creativity) without in so doing making creative effort(i.e. without the process of creation). One and the same task (teaching or in production) is solved in different cities by different people using the very same rules and identical results are obtained independently of the individual qualities of the people involved. This result (invention) is formally called creative, but in fact it is a common or gardensuch as, for instance the calculation of a beam according to the formulae for the resistance of materials.

This paradox has been caused by the fact that the concept of "creativity" is not somet6hing unchanging and frozen: the contents invested in this concepts are constantly changing. In the Middle Ages, for example, the solution of equations of the third degree was genuinely creative. Tournaments were organised in which mathematicians challenged each other with equations, or square root problems. But then the Cardano formula came out and the solution of third degree equations was open to every first year mathematician.

Now imagine a "transitional period"; everyone is looking for the roots of algebraic equations, eliminating possibilities, but we both know the Cardano formula. They all consider us geniuses - or at least very talented - but we know that it is the formula at work.

TSIP enables us today to solve inventive problems on the level of organization of mental activity that will be the norm tomorrow.

When one and the same problem is solved by two persons, one using elimination of possibilities and the other using TSIP it is like a race between a runner and a car driver. One moves on his own feet, as it were, the other is speeded along by a powerful engine, but the judges are only interested in the time taken. Today TSIP is like a car at the beginning of this century: a new machine, still far from perfect but obviously far faster than a man and, most important, open to almost limitless improvement in the future. TSIP has so far not mastered certain classes of problem (obtaining new substances, elucidating optimal work rates, etc). With time these problems too will come within the power of TSIP since there are no fundamental objections to it.

The reader will rightly be asking, does this mean that the moment will come when all inventions will be carried out "according to formula" and invention will cease to be a form of creative activity? Yes, that day will come. Inventive creativity - all this trial and error, "enlightenment", "happy coincidences" are not an end in themselves but a means of developing technical systems. The means is so imperfect that 17 centuries ago the idea was already expressed of the need to replace "creativity" by a more effective method - "science". One has returned to this idea on many occasions. But until recent times there was no need to invent in science - one simply increased the number of inventors. Now the situation has changed. It has become far more difficult to "pick a number", and the waste of time necessary for using the trial and error method has become inadmissible. The appearance of TSIP, and its rapid development is not an accident but a necessity, produced by the contemporary scientific and technological revolution.

The design of technical systems which one hundred years ago was an art has become an exact science today. Until recent times this science included only the design of well-known technical systems and did not touch on the creation of fundamentally new ones. Now one speaks with assurance of the fact that the design of systems has turned into the science of the development and design of technical systems. Invention as a method of creating new systems and perfecting old ones is historically played out. It is demeaning and unreasonable to "dig by hand" where a machine can be used. Work "by formula" inevitably drives out work "by feel". But the human mind will not remain without work. People will think over more complicated problems.

The replacement of the science of invention is a complex and unhurried process, dependent not only on the development of the TSIP but also on the evolution of the theory and practice of the patent protection of inventions, in first place from the gradual alteration of the contents included in the concept "invention". As we have seen, the lower threshhold of the requirement for a technical solution which would claim to be an invention is now extremely low. Even the most trivial proposals are frequently patented as inventions. One must suppose that in the foreseeable future requirements for inventions will be raised. A few years ago the handbook on patent applications for the first time contained the idea that an invention is the removal of a technical contradiction: "Thus a necessary condition for the appearance of an invention is the presence of a contradiction inherent to known solutions of a technical problem. In order to test for the presence of a invention in a specific proposal it is required that this contradiciton be revealed in known solutions of problems and establish that the given proposal permits one to remove this contradiction in part or in full."[29,p.20]. So far this is what has been written in the handbook; the introduction of a similar definition in normative acts on invention has led to the fact that present inventions of the first and second levels have ceased being considered inventions. Specifications for inventions should be regularly reviewed and raised.

Very much also depends on the system of instruction of TSIP. So far this system is not great but it is quickly growing and its very existence creates the pre-conditions for a transition in the future to mass instruction; experience in instruction is being accumulated and textual and visual aid materials are being created, the training of lecturers is proceeding.

Instruction in TSIP is being organized by various ministries and departments, the administration of enterprises, the scientific-research institutes, universities and colleges, Communist Youth League area and city committees, the All-Union "Knowledge" Society, the Scientific-Technical Society, the All-Union Society of Inventors and Rationalizers. Studies are organized on three levels:

1. An informative series of lectures (20 teaching hours). The aim of such lectures is to present the basic principles of TSIP, to show the necessity for serious study of the theory. The lectures allow one to select students for classes at a school of inventine creativity.

2. School (100-120 teaching hours). Clases are usually held once a week. The aim is to teach the techniques of the application of ASIP. The school programme corresponds to that of the first year of a public institute of inventive creativity. Some students go on to continue their studies at the institute.

3. Institute (220-240 teaching hours). The aim is to train specialists in TSIP (elaborators, lecturers).

Recently TSIP has become a teaching subject at certain institutes for raising the qualifications of managerial personnel. Classes are organized in a programme of 144 hours and held (as a release from production work) over a month. Then students continue with self-studies. In 5 to 6 months they are examined in their project work. Each project consists of solving a problem of direct relevance to industry on invention level and the appropriate methodological analysis of the process of solution.

How effective are these studies?

The press has on various occasions published reports on inventions made by graduates of public schools and institutes [19.pp 30-33]. For instance, in the newspaper Magnitogorskiy metall for 24 and 26 April the engineer M.I.Sharapov described how an "obsolete" problem was solved using ASIP. The inventions emerging from this solution were given Pat. Cert. Nos 212 672 and 239 759. Their implementation in the Magnitogorsk Combine alone saved 42,000 rubles annually (Izobretatel' i ratsionalizator, 1974, No 1, p.24). M.I.Sharapov now has over 40 patents and almost all inventions have been implemented.

"For over ten years researchers at our designers bureau tried to make a simple and reliable system of programmed control", wrote the inventor Yu.Chinnov in the article The Logic of Success [34], "and I and my many colleagues were met only by failure. In 1967 specially in order to check the methodology of inventive creativity I chose this problem since I considered it to be sufficiently complex and I was not certain that the method would help me solve it. But first of all the question was one of testing the method. The problem of construction of a fundamentally new, reliable and simple system was solved" (Pat. Cert. Nos 222491 and 248 819). Yu.V.Chinnov, Honoured Inventor of the Uzbek SSR now has around 70 patents, more than 50 of which were awarded for inventions made with ASIP.

Many such examples can be cited. One would think, however, that it is more important for all graduates of public institutes (and in recent years from public schools too) to complete their studies with diploma works on inventor level. As "Pravda" wrote on May 6, 1975: "Three years ago a score or so of young people became students at the country's first institute of inventive creativity, set up under the Central Committee of the Communist Youth League of Azerbaijan and the Republican Council of the All-Union Society of Inventors and Rationalizers. Before this they had not displayed any special aptitudes for technical creativity, and the institute selected students without reservations and registered everyone who applied. They emerged from the institute as fully fledged inventors: some of them with patents, the others with glittering creative prospects, which can be judged from the excellent appraisals given to the majority of students for their diploma works. Graduates of the institute solved real technical problems which had so far resisted the efforts of inventors."

One should note that studying at schools and institutes of inventive creativity are not only engineers but also educationalists, physicians, mathematicians, chemists, etc. As we have already said, the TSIP develops systematic thought, and the ability to think in an organized fashion, to control the process of thought is necessary not only in technology. The classes are conducted on technical matters, but the mind develops "in general". And this is entirely natural. If a man examined 200 to 300 technical problems and trained himself to see each inventive situation on several levels (sub-system, system, super-system), if he trained himself to feel out the dialectic of development (the origin and overcoming of contradictions, the unity of opposites in the object-antiobject), to operate with concentrated information and see in a particular problem general laws of development, such a style of thinking cannot but spread to beyond the bounds of technical systems. Just as in sport where a man takes up one definite sport but the "spin-off" is seen in a general strengthening of the organism.

The characteristic feature of TSIP studies is that students not only receive ready-made knowledge, but also participate actively in the process of production. From the first days of his studies each student begins to maintain a personal card index, gathering information about new methods, physical effects, materials and especially successful and elegant technical solutions, etc. Then from their personal card indexes (and there are thousands of entries) the most interesting ones are selected for inclusion in the overall card index, maintained in conjunction by all schools and institutes. It brings together all the most interesting things, and in it the most diverse streams of information collide and mutually interact.

There is one other feature of no small importance. The method of trial and error is individualistic by its very nature; each one selects variants; one can "fragment" a problem but never subdivide labour. This individuality is retained even in the brain storm and in synectics; the same cottage industry production of ideas, only that it is a team of cottage artisans. The TSIP creates the preconditions for the sharing of creative labour, nd by so doing lays the foundations for the transition to genuinely collective creativity. Even in the process of study are revealed "the partisanship" of students to this or that school of theory. In the course of time such "partisanship" is converted into a kind of specialisation.

The possibility emerges of systematically sharing work within the team; of forecasting the themes of inventions and revealing new problems, the regular stocking up of the information pool independently of current needs, the analysis of problems, the use of DTC (Dimension-Time-Cost) operators and other special psychological methods, converting ideas received into the fund of ideas (Step 6,3). At the same time with specialization like this the whole team retains a common language (IFR, TC, PC, etc) and a common methodological approach to the problem.

THERE ON THE HORIZON

And so, inventions, even those of the very highest level can also be made "according to formula". Many of these "formulae" are already known and their application is studied with success. In the next decades the solution of inventive problems will be converted into an exact science on the development of technical systems.

At this point what we know for a fact comes to an end. In order to look in the future one needs to enter the region of supposition and conjecture but all the same, I will risk expressing a few thoughts.

Evidently the reader has already long ago asked, "Well, this is all right, technical systems will develop 'according to formulae' without the 'torments of creativity' but what about science and discovery?"

Technology is "populated" with developing systems, machines,, whereas science has developing systems and theories. The life of theories is subordinate to laws which in many ways coincide with the laws of the development of machines. They are the same in nature and the difficulties arising in perfecting both systems by the trial and error method. Here is a typical example: In G. Watson's book <u>The Double Helix</u> it relates how one hypothesis on the structure and mechanics of the reproduction of DNA arose. Francis Crick, a future Nobel Prizewinner heard pver beer about an astronomist who had talked something about the "perfect cosmological principle." Crick paid little attention to exactly what he meant by this but he was set to wandering whether an argument could be made for the 'perfect biological principle', i.e. could one formulate the IFS for the problem under consideration? Crick was interested in the problem of the replication of DNA and therefore in this instance the "perfect biological principle" was "the self-replication of DNA." That is how the first step toward one of the most important discoveries of the 20th century was taken.

An accidental phrase heard in a university pub... But what if one had consciously employed the IFS for the "making of a discovery" if only in biology itself? Take Engineer (not a biologist!) G.G.Golovchenko, a TSIP lecturer from Sverdlovsk, who applies the IFS concept to .... the evolution of plants. His course of thinking was approximately as follows: plants of today in comparison with those of yesterday are an approximation to the IFS, to some ideal plant which has absorbed to the full substances and energy from its habitat, and let us look at what contradictions could arise in this process and how the plant overcome them, in drawing closer to the IFS. The research undertaken by Golovchenko led to the discovery of the wind energy power of plants - the ability of plants to directly utilise the energy of the wind.[35].

The DNA molecule, a gigantic double helix, upon reproduction at first divides into two separate spirals. Aminoacids attach to each of them and form two double helixes. For a long time there was no explanation of how the double helix managed to untwist. Calculations showed that the DNA of intestinal bacteria should untwist at a speed of 15,000 revolutions a minute for otherwise DNA would not manage to replicate itself in the time in which it has been observed to do so. Only recently has one been able to establish where the problem is. One of the spirals breaks up in many places; each section of the helix untwists of itself and hence only a few turns are needed and the section is unravelled. Later they join up once more into an integral chain.

And so, there is a contradiction: the molecule should untwist and the molecule should not untwist. The method of overcoming it is :dividing and rejoining.It is interesting to compare the idea of this scientific discovery with that of the invention made by Yu.V.Chinnov [30]. To make a cable one needs to coil wires. For this purpose the wires are fed into a coiling frame. As the frame rotates the wires are twisted and have joined together as a cable by the time they reach the take-up spool inside the frame. The question is, why should the spool be placed inside the frame. This only causes trouble, since it has to be made small and the cable obtained has to be in sections; upon rotation of the spool considerable centrifugal forces are set up which limits the productivity of the rig. However, it is impossible to locate the spool outside the frame: if the cable is twisted "in the passage" (i.e., onto a spool located outside the frame) one obtains a "false coil". For the amount that is twisted before the frame the same amount will untwist after leaving the frame. The specialists warned Chinnov that inventing a method of twisting cable is every bit as impossible as inventing a perpetual motion machine.

Chinnov utilized ASIP-68 and solved this problem. The idea of the solution is that molten metal enters the rotating frame which hardens before it has left it. Melting is also a process of dividing but at micro-level. The contradictions in the problems (scientific and inventive) are the same: accordingly the solutions are the same.

Making an invention means thinking up a technical system that has none of the contradictions present in its antecedent system. It is exactly the same thing with a discovery. This means thinking up a theory (or scientific system) which has no contradictions inherent to preceding theories.

At the beginning of the 20th century the so-called Russell Effect was discovered. It transpired that certain metals, if their surface is cleansed of its oxidized surface will give an image onto a photosensitive film applied to it in darkness. Many researchers were interested in this discovery. It was established that hydrogen atoms formed from the interaction of metal and water vapour act on the surface. A clear cut case, on would think. But suddenly something incomprehensible turned up. The film also turned black whenever it was placed ten or more millimetres away from the metal, although the hydrogen atoms could not possibly leap such a distance. For seventy years the paradox was not susceptible to explanation. The Leningrad engineer and TSIP lecturer V.V.Mitropanov used the concept of physical contradiction to explain this paradox. The hydrogen atoms have to travel a great distance in order to act upon the film; hydrogen atoms may not travel a great distance since they would "perish" in so doing. The contradiction is very similar to the one we encountered in Problem 61 on plating (a metal should and should not be in a lubricant). The solution is similar too (standard 10). The hydrogen atoms forming on the surface of the metal are joined onto the excited molecules: such molecules last the "leap" to the film but having done so proceed to decay under the effect of surface forces. The hydrogen atoms are there and yet they are not... The paradox had been cracked [36].

Now compare the invention of the method of measuring the mobility of ions (Problem 35) with the discovery of the nature of the Russell Effect. The same physical contradictions, similar methods of overcoming them. The difference is that in the first instance the Idea has been converted into a Thing and in the second in a Theory. The forms of implementation (instilling) are different, but the mechanics of the solution are the same.

The explanation of the Tunguss Explosion is a typical "problem of discovery". One needs to conceive of a cosmic body which would produce a strong localized explosion (but not nuclear in nature, since there is no trace of a nuclear explosion); in order to obtain a localised effect one has to close down the dimensions of the supposed object but then one could not explain the force of the explosion. The mass is small, the speed relatively small (according to the evidence of eye-witnesses) so where did the energy come from? One needs to construct a model with no contradictions - and working on this level there is no difference between a discovery and an invention.

Certain researchers including G.L.Filkovsky consider that practically the whole TSIP apparatus should be used to construct a theory of the development of technical systems. time will tell.

It is no accident that the revolution in methods of problem solving began in technology. Only there is there a pool of patents. In science and art information on innovations is dispersed and dissipated in disparate literature. The laws of the development of systems are revealed more clearly in technology: an unsuitable theory sometimes goes on living for a long time, but an unsuitable machine simply will not work. An inventor can seek after readymade keys to problems in physics, and where can he look for such keys (and do they even exist?) for a "problem of discovery?"

Yet all the same the new technology of the solution of creative problems in this form or the other inevitably spreads beyond the limits of technology. How far is difficult to say. Who could have foretold that from Hertz's experiments, Maxwell's Equations and Popov's "thunder marker" radio technology would emerge which now in one way another impinges on the life of everyone person on earth?

Evidently the possibilities of controlling the process of thought are boundless. They cannot be exhausted, because Reason, the greatest instrument of knowing and transforming the world, is also capable of transforming itself. Who can say that that there is a limit to the processof humanization of man? As long as Man exists the control of this force will be improved. We are only at the start of a long road.

PROBLEMS

The majority of problems cited in preceding chapters have been examined. In order to give the reader the opportunity to train himself we offer a few more instructional problems. Remember: solving a problem means first of all establishing the rule on the basis of which the task will be solved and only then giving a specific answer.

PROBLEM 62

Parts made of rubber and other elastic materials cannot be worked on a lathe. How can one work a material which changes its shape at the slightest pressure from the cutter? The Czechoslovak inventor I. Pauker has solved this problem (Patent of the CSSR No 121 621). What is the idea of the invention? What standard should be used for solving tasks of this kind?

You can check your answer in the journal Izobretatel' i ratsionalizator, 1969, No. 2, p.27.

PROBLEM 63

The Toms Effect was mentioned in Chapter 7 according to which the addition of a small quantity (a fraction of a percent) of long-chain polymers to a liquid has the effect of considerably lowering the friction of the liquid moving past a solid surface, the walls of a pipe, for instance. But what if these polymers are added to a solid body? What solid body could take such additives and what would be the effect?

For the answer see the bulletin Otkrytiya, Izobreteniya, Promyshlennye obraztsy. Tovarnye znaki, 1974, No.35, p.88, PCN 444 039.

PROBLEM 64

Car fuel tank indicators are inaccurate and capricious. A case arose where the indicator said the tank still contained a lot of petrol whereas in fact there was almost nothing left; the car went out onto the highway then stopped. One needs a very simple method of informing the driver that the minimum quantity of petrol is in the tank and that it is time to fill up.

Remember that there already is a floating sender of electrical information. Something else is needed.

Solve this problem making use of the table of typical models and S-Field transformation. If your solution can be installed by every driver in five minutes without any remodelling of the equipment and also that the equipment does not cost more than 20 to 30 kopeks and the method is absolutely reliable this means you have solved the problem correctly.

The answer is to be found in Pionerskaya pravda for 15 November, 1977.

PROBLEM 65

A laboratory has a powder-like oxide of beryllium whose melting point is over 2000C. One needs to melt the beryllium oxide in such a way that it is not contaminated. There is no crucible in the laboratory that would stand up to the necessary temperature. The idea has arisen of melting the beryllium oxide within the beryliium oxide itself. Let us take a "pile" of beryllium oxide and heat the middle of it with a high frequency current. The melted portion should not come into contact with anything (apart from beryllium oxide) and therefore is not polluted. All is fine, but the trouble is that beryllium oxide starts to conduct electricity only at a high temperature. How can it be heated up? An attempt was made to use an electric arc, a plasma lance induction heating, but without success. The beryllium oxide either did not heat up or gets dirty. There is a clear contradiction: in order to make beryllium oxide an electroconductor, one must introduce metal into it but in order to preserve its purity metal cannot be introduced.

Solve this problem using the standards. The answer can be checked in the journal Izobretatel' i ratsionalizator, 1977, No. 3, p.23.

PROBLEM 66

At an engine factory, after assembly the engines are sent for running in. Here the drive of the engine is coupled to an electric power unit which turns at a constant relatively small number of revolutions. The pistons of the engine come into motion relative to the internal surfaces of the cylinders and are gradually ground down. The unevennesses, prominences, rough spots are smoothed away and the pistons now fit the cylinder walls better. In essence the process is beautifully simple: one rough surface grinds against another until all roughness has been smoothed away.

The running in process continues until the pistons no longer grind against the cylinder walls. But when is that moment? Attempting to monitor the process by introducing luminescent traces into the oil and following the rate at which luminescence disappears as metal particles enter the oil have proved to be too unwieldy. Even more unwieldy was the method of periodically stopping the machine to take it apart and examine the rubbing surfaces.

Solve this problem using ASIP from step

2.2. The answer can   be found in the bulletin Otkrytye. Izobreteniya. Promyshlennye obraztsy. Tovarnye znaki, 1972, No. 15, p. 155 and PCN 337 682.

PROBLEM 67

D.Peev's   story   "The   Seventh   Goblet" (Iskatel', 1975, Nos. 1 and 2) describes the investigation of a murder that took place in mysterious circumstances. Seven people (the case occurred in a country house) were gathered for a dinner party. Brandy poured into glasses was not drunk immediately but the guests toured the house for the next fifteen minutes. Then they returned to the table and drank the brandy. One fell dead - there was poison in his brandy glass. The investigator established that in the course of that 15 minutes no one had been out of sight of the others, and no one could have poured the poison in.

How is the crime unravelled?

PROBLEM 68

For a long time porous bodies have been examined by taking a section and studying it under the microscope (for instance, to study the shape and disposition of pores).

What has to be done to perfect this method?

If this problem causes you any problems re-read Chapter 2.

284 CREATIVITY AS AN EXACT SCIENCE

For the answer see the bulletin Otkrytie. Promyshlennye obraztsy, 1974, No. 32, p. 103, PCN 441 481.

PROBLEM 69

Capillary action helps solder to penetrate into hardly perceptible gaps between parts. That very same action is harmful, however, when it is necessary to solder a porous insert into the inner surface of a hub. The solder penetrates the pores and blocks them. What can be done?

For the answer see Izobretatel' i ratsionalizator, 1977, No. 3, p.44.

PROBLEM 70

Polymers are constantly spoiling and growing old. This happens because free radicals arise from the action of oxygen in the polymer. In order to protect the polymer it is necessary to introduct substances into it that will intercept the oxygen, such as highly ground metals. But how can such a metal be introduced without it acquiring a coating of oxide which would deprive it of it "intercepting" properties? Finely ground metal avidly unites with oxygen and carrying out the "correction" to the polymer in a vacuum or in an inert gas medium is too complicated.

What standard should be employed to solve this problem? what should be done in accordance with this standard?

APPENDIX 1

ALGORITHM FOR THE SOLUTION OF INVENTIVE PROBLEMS ASIP-77

Part 1: Selection of Problem

1.1. Determining the final aim of the solution of a problem:

a. What features of the object need to be changed?

b. What features of the object is it known cannot be changed in solving the problem?

c. What costs can be lowered if the problem is solved?

d. What (approximately) is the permissible expenditure?

e. What is the main technical-economic indicator that should be improved?

1.2. Checking the deviation route. Let us suppose that the problem is fundamentally insoluble. What other problem must be solved in order to obtain the requisite final result?

a. Transform the problem by switching to supersystem level which the system given in the problem would enter.

b. Transform the problem by switching to subsystem level (of substances) included in the system given in the problem.

c. On three levels (supersystem, system and subsystem) transform the problem, having replaced the requisite action (or property) by its reverse.

1.3. Determining the solution of which problem is more purposeful - the original one or one of the deviations. Making a choice.

NB. In choosing one should take account of factors which are objective (what are the reserves of development of the

285

system given in the problem) and subjective (is one orientated toward which problem - the maximum or the minimum).

1.4. Determining the requisite quantitative indicators.

1.5. Increasing the requisite quantitative indicators, taking account of the time necessary for implementing the invention.

1.6. Specifying exactly the requirements necessitated by the specific conditions in which the realization of the invention is proposed.

a. Taking account of the features of implementation, in particular the admissible degree of complexity of the solution. b. Taking account of the proposed scale of implementation.

1.7. Checking whether the solution is solved by a direct application of standards on the solution of inventive problems. If an answer is obtained, go on to 5.1. If not, go on to 1.8.

1.8. Specifying the problem by using patent information.

a. What answers can be drawn from patent information on problems which are related to the one in question.

b. What are the answers to problems similar to the one in question but relating to the leading branch of technology.

c. What are the answers to problems which are the reverse of the one given.

1.9. Using the DTC operator.

a. By mentally changing the dimensions of the object from its real size to 0 how can the problem be solved now?

b. By mentally changing the time of the process (or speed of movement of the object) from that given to how can the problem be solved now?

c. By mentally changing the cost (allowable expenditure) of the object of process from that given to 0 how can the problem be solved now?

Part 2: Constructing a Model of the Problem

2.1. Noting down the specifications of the problem with using special terminology.

Examples

(Problem 24)
A polishing disk is poor at processing a product with a complicated shape with recesses or protuberances, such as a spoon. Polishing in some other way is inconvenient and complicated. Using polishing disks made of ice is too expensive in this case. Also unsuitable are elastic inflatable disks with an abrasive surface since they wear down too quickly. What can be done?

(Problem 25)
The aerial of a radio telescope stands in a locality of frequent storms. In order to protect it from lightning it is necessary to erect lightning conductors (metal rods) around it. But lightning rods keep back the radio waves and set up a radio shadow. Erecting the lightning conductors on the antenna itself is impossible in this case. What can be done?

2.2. Isolate and write down the conflicting pair of elements. If in the specifications of the problem only one element is given, go on to step 4.2.

Rule 1. It is necessary to introduce an artefact into the conflicting pair of elements.

Rule 2. The second element in the pair should be one on which the artefact (an instrument or second artefact) can act directly.

Rule 3. If one element (instrument) according to the specifications of the problem can be in two states then one should choose that state which ensures the best implementation of the main productive process (the basic function of the whole technical system indicated in the problem).

Rule 4. If the problem has a pair of homogeneous mutually interacting elements (A1, A2... and B1, B2...) it is sufficient to take one pair only (A1 and B1).

Examples

The artefact is a spoon. The instrument is a polishing disk acting directly on the artefact.

In the problem there are two "artefacts" - lightning and radio waves - and one "instrument" - a lightning conductor. The conflict in this instance is not within the "lightning conductor - lightning pair" or the "lightning conductor - radio waves" but between these pairs.

In order to transpose such a problem into canonical form with one conflicting pair it is necessary in advance to give the instrument a quality necessary for fulfilling the basic productive operation of the given technical system, i.e., one needs to take it that there is no lightning conductor and that the radio waves freely pass through to the antenna.

Thus the conflicting pair is: an absent lightning conductor and lightning (or a non-conducting lightning rod and lightning).

2.3. Writing down two interactions (actions, properties) of elements of the conflicting pair which have what one needs to introduce; useful and harmful.

Examples

1. The disk possesses the ability to polish.

2. The disk does not possess the ability to adapt itself to curved surfaces.

1. An absent lightning conductor does not create radio interference.

2. An absent lightning conductor does not attract lightning.

2.4. Writing down the standard formula of the model of a problem, having indicated the conflicting pair and the technical contradiction.

Examples

A disk and an artefact are given. The disk possess the ability to polish but cannot adapt itself to the curved surface of the artefact.

An absent lightning conductor and lightning are given. Such a lightning conductor does not set up radio interference, but neither does it attract lightning.

Part 3. Analysis of the Model of the Problem

3.1. Selecting from the elements in the model of the

problem the one that can easily be changed, replaced, etc.

   Rule 5. Technical objects are easier to change than
natural ones.
   Rule 6. Instruments are easier to change than artefacts.

   Rule 7. If in the system there are no easily changeable
elements, one should indicate an "external environment".

Examples

The shape of an artefact cannot be changed: a flat spoon
would not hold liquid. The disk can be changed (while
retaining, however, its ability to polish - these are the
specifications of the problem).A lightning conductor is an
instrument for "processing" (changing the direction of
movement of) lightning which in the given instance is
considered as an artefact. The analogy is with a rainpipe
and rain. Lightning is a natural object and a lightning
conductor is a technical object, and so the one to choose
is the lightning conductor.
3.2. Noting down the standard formula of the IFR (the
ideal final result).
An element (indicate an element chosen at step 3.1) itself
removes the offending interaction while retaining the
ability to perform (indicate the useful interaction).
   Rule 8. The IFR formulation should always include the
word "itself".

Examples

The disk itself adapts to the curvature of the artefact while retaining the ability to polish.

The absent lightning conductor itself ensures the "catching" of the lightning, while retaining the ability not to set up radio interference.

3.3. Identifying' that zone of the elment (indicated in step 3.2) which cannot cope with a set of two interactions demanded in the IFR. What is there in this zone, a substance, a field? Show this zone on a schematic drawing denoting it in colour, lines, etc.

Examples

The external layer of a disk (the outer ring, frame); a substance (abrasive powder, solid body).

That part of space which is taken up by the absent lightning conductor. The substance (a column of air) freely penetrated by radio waves.

3.4. Formulating contradictory physical requirements made of the state of the isolated zone of the element of the conflicting interactions (actions, properties).

a. For security (indicate the useful interaction or that interaction which should be retained) it is necessary (indicate the physical state: being heated, mobile, charged, etc);

b. For prevention (indicate the harmful interaction or interaction that should be introduced) it is necessary (indicate the physical state: being cold, immobile, uncharged, etc).

Rule 9. The physical state indicated in paragraphs a and b should be mutually opposed.

Examples

a. In order to polish, the outer layer of a disk should be firm (or rigidly fixed to the central part of a disk for transmission of force).

b. In order to adapt to the curvature of an artefacts the outer layer of the disk should not be firm (or should not be rigidly fixed to the central part of a disk).

a. In order to let through radio waves a column of air should not a conductor (or more precisely, should not have free charges).

b. In order to attract lightning the column should be a conductor (or more precisely it should have free charges).

3.5. Noting down the standard formulation of the physical contradiction.

a. The full formulation: (indicate the denoted zone of the element should (indicate the state noted at step 3.4a), in order to perform (indicate the useful interaction) and should (indicate the state noted in step 3.4b) in order to prevent (indicate the harmful interaction).

b. The short formulation: (indicate the denoted zone of the element) should be and should not be.

Examples

a. The outer layer of a disk should be firm in order to polish an artefact, and should not be firm in order to adapt to the curvature of the artefact.

b. The outer layer of the disk should be and should not

be.

a. A column of air should have free charges in order to "catch" the lightning and should not have free charges in order not to trap the radio waves.

b. A column of air with free charges should and should not be.

Part 4. Removing the Physical Contradiction

4.1. Examining the simplest transformations of the denoted zone of the element, i.e., separating the contradictory properties

a) in space;

b) in time;

c) by using transitory states in which contradictory properties either co-exist or appear alternately;

d) by rebuilding the structure: particles of the denoted zone of the element are given which possess the property, and all the denoted zone as a whole are marked by the requisite (conflicting) property.

If a physical answer is obtained (i.e. it emerges that a physical action is necessary) go to 4.5. If there is no physical answer, go to 4.2.

Examples

Standard transformations do not produce an obvious solution to Problem 24 (although, as we shall see later, the answer is close to 4.1 c and e).

Problem 25 can be solved by 4.1 b and c. The free charges themselves appear in the column of air at the initial stages of the origin of lightning. The lightning conductor for a short moment becomes a conductor and then the free charges themselves disappear.

4.2. Using the table of typical models of problems and S-Field transformations. If a physical effect is obtained go to 4.4. If there is no physical effect switch to 4.3.

Examples

The model of Problem 24 belongs to class 4. According to the typical solution the substance S2 should be developed into an S-Field, introducing the field F and adding S3 or dividing S2 into two interacting parts. (The idea of dividing the disk was beginning to take shape in at step 3.3. But if one simply divides the disk the outer part will fly off under the effect of centrifugal force. The central part of the disk should hold on strongly to the outer part and at the same time should give it the possibility of being freely exchanged...) Later according to the typical solution it is desirable to translate the S-Field (obtained from S2) into an F-Field, i.e. to use a magnetic field and a ferromagnetic powder. (This enables one to make the outer part of the disk movable, exchangeable and guarantees the requisite bond between both parts of the disk).

The model of problem 25 belongs to class 16. According to the typical solution the substance S1 should be doubled, becoming either S1 or S2, i.e., the column of air should become conducting on the appearance of lightning and then turn into a non-conducting state.

4.3. Using the table of application of physical effects and phenomena. If a physical effect has been obtained go to 4.5. If there is no physical effect then 4.4.

Examples

Problem 24: from the table item 17 is suitable - the substitution for "substance" of "field" bonds by means of using electromagnetic fields.

Problem 25: from the table item 23 is suitable - ionization under the effect of a strong magnetic field (lightning) and their reconstitution after the disappearance of this field (radio waves are only a weak field). Other effects relate to liquids and solid bodies, and call for the introduction of additives or do not ensure automatic self-regulation.

4.4. Using the table of basic methods of eliminating technical contradictions. If a physical effect has been obtained before this use the table to check it.

Examples

According to the specifications of Problem 24 one needs to improve the ability of the disk to "grind" artefacts of different shapes. This is adaptation (line 35 in the table). One way is to use a selection of various disks. On the debit side one loses time when exchanging and selecting disks and lowering productivity: columns 25 and 39. The methods according to the table are 35,28; 35,28,6,37. The recurring and hence most likely methods are 35: exchanging the total state (the outer part of the disk is "pseudoliquid, made of moving particles); and 28 - a direct indication to switch to an F-Field, which has been done above.

According to the specifications of Problem 25 one needs to eliminate the action of lightning - a harmful external factor (line 30). One way is to set up the usual metal lightning conductor. On the debit side a radio shadow is set up i.e., a harmful factor is created by the lightning rod itself (column 31). In the table this box is empty. Let us take column 18 (reducing illumination, the appearance of an optical instead of a radio shadow). Methods: 1,19,2,13. By Method 19 one action is carried out in the pauses of the other.

4.5. Switching from a physical answer to a technological: formulating a method and giving a design of a device which would implement this method.

Examples

The central part of the disk is made from magnets. The outer layer of ferromagnetic particles or abrasive particles baked onto the ferromagnetic particles. Such an outer layer will assume the shape of the artefact. At the same time it will retain its firmness necessary for polishing.

In order that free charges appear in the air it is necessary to reduce pressure. A casing is needed to keep in this column of air under reduced pressure. The casing can be made of dielectrics for otherwise it would itself set up a radio shadow.

Pat. Cert. No 177 497: "A lightning conductor which, in order to have the property of radio transparency, is made in the shape of a hermetically sealed pipe made of a dielectrical material the pressure within which is selected from the condition of the least as charged gradients caused by the electrical field developed by lightning."

Part 5: Preliminary Assessment of Solution Obtained
5.1. Carrying out a preliminary assessment of the solution obtained.

Checklist
1. Does the solution obtain ensure the fulfillment of the main requirement of the IFR ("The element itself...)
2. Which physical contradiction has been eliminated (if at all) by the solution obtain?
3. Does the system so obtained contain at least one well controllable element? Which? How is this control implemented?
4. Is the solution found for a "one-off" model of the problem also suitable in the real world using seriel production?
   If the solution does not satisfy at least one of the checklist questions return to 2.1.
5.2. Check (against patent information) whether the solution obtained is formally an innovation.
5.3. What side effects could could arise in the technical implementation of the idea so produced? Note down possible side problems inventive, design, accounting, administrative.

Part 6. The Development of the Answer Obtained
6.1. Determine how the super-system to which the changed

system belongs will itself have to changed.

6.2. Check whether the changed system will have to be applied in a new way.

6.3. Make use of the answer so obtained in the solution of other technical problems.

a. Examine the possibility of using the idea which is the reverse of the one obtained.

b. Construct a table on "The disposition of parts in the overall state of the artefact" or on "The use of the field in the overall states of the artefact" and examine possible reconstruction of the answer seen in the light of these tables.

Part 7: Analysis of the Process of Solution

7.1. Compare the real course of the solution with the theoretical (according to ASIP). If there are any deviations, make a note of them.

7.2. Compare the answer obtained with tabular data (the table of S-Field Transformations, the table of physical effects, the table of basic methods). If there are any deviations, make a note of them.

APPENDIX 2

TYPICAL MODELS OF INVENTIVE PROBLEMS AND THEIR S-FIELD TRANSFORMATIONS

Type 1. One element is given

1. The substance lends itself poorly to control (inspection, measurement, change); effective control must be ensured.

a. The general way of solving problems of this class is the completion of an S-Field (the introduction of a second substance and field).

b. For problems of inspection and measurement use standard 1. The introduction of a second substance (a luminescent trace, a ferromagnet etc), interacting with an external electromagnetic field):

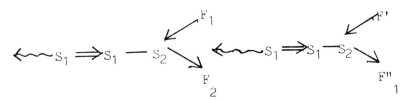

c. For problems of relocation, subdivision, treatment of the surface, deformation, changing viscosity, solidity, etc, apply standard 4. The introduction of ferrmomagnetic particles and a magnetic field:

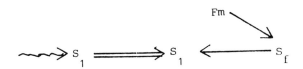

299

d. If it is impossible to introduct S2 apply standard 8
(measuring the inherent frequency of vibration) and 10
(roundabout routes: instead of S2 introduce a field and
also an "external" S2, S2 is introduced temporarily or in
extremely small doses, use as S2 part of Sl, se instead of
an object a copy of it, introduce S2 in the form of
chemical compounds of it).

2. The field is badly susceptible to control (inspection,
measurement, change, transformation into another field);
effective control is mandatory.

a. Transformation of the original field Fl with the help
of a converter substance or two interacting substances:

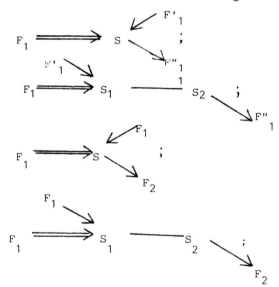

b. Introduction of substance Sl which changes its
properties under the action of field Fl, and in this
process this change is easily revealed with the help of
field P2 acting on S:

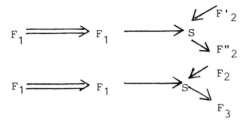

3.   The substance (or field) possesses two attendant conflicting properties: one property should be improved without spoiling the other.

a.   Problems of this class are translated into Class 1 ann 2 by replacing the original substance S (or field F) by substance S' (or field F') to which earlier one of the attendant properties has been has been applied in full measure:

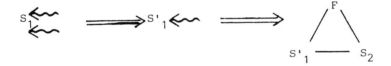

For instance, the problem "The neight of an antenna should be increased without increasing its weight" is transposed into "A high antenna should be as light as a low one." Of the two attendant properties one should ascribe to the object in advance the one that ensures the maximum efficiency of the basic action. Therefore one takes a high antenna and not a low one.

b.   If the property conflicts with its antithesis (hot - cold, strong - weak, magnetic - non-magnetic) then it should be overcome by separation in space, time and structure (the whole has one property, a part has another). If one uses separation of substances in time it is expedient that the transition from one state to another be carried out by the substance itself, taking on different forms in turn (changing the aggregate state, moving through the Curie Point, dissociation - association, etc).

Type 2. Two elements are given.

4. Two substances do not interact (or interact very badly); one substance (or both) can be changed: the requirement is to ensure a good interaction.
Substance S2 is converted into an S-Field, which forms a chain with S1 and in this way ensures the interaction of S1 and S2; then, if possible, the S-Field converted from S2 is translated into an F-Field, i.e., an S-field with a magnetic field Fm and the ferromagnetic substance Sf. (preferably in the form of tiny particles);

$$S_1 - - -S_2 \Longrightarrow S_1 \underline{\quad} S_2 \overset{F}{\triangle} S_3 \Longrightarrow S_1 \underline{\quad} S_2 \overset{F_m}{\triangle} \underline{\quad} S_f$$

$$S_1 - - -S_2 \Longrightarrow S_1 \underline{\quad} \triangle \underline{\quad} S_2$$

If, despite the fact that S2 has been converted into an S-Field a direct connection is not established between S2 and S1 one can use the connection via field F:

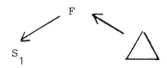

In problems of measuring or revealing S2 is converted into an S-Field with the field at the exit, for instance:

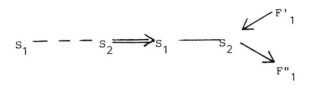

5. The same as in class 4 but both substances cannot be changed.

a. The problem is translated to class 4 using standard 10.

b. Instead of substances one uses optical copies of them.

6. Field Fl is not controlled by field F2: the requirement is to ensure effective control.

The introduction of substances (or two mutually interconnected substances), whose ability to interact with P2 does not depend on the action of Fl:

$$F_1 \,\text{---}\, \text{--} \, F_2 \Longrightarrow F_1 \,\text{---}\, S \,\nearrow^{F'_2}_{\searrow_{F''_2}}$$

The degree of control in some instances can be increased by dint of using a substance such as S which under the action of Pl undergoes a phase conversion (for instance, melts, passes through the Curie Point, etc).

7. The field and substance do not interact; their interaction is required.

Introduction of a mediator substance S2 or a battery of

substances (S2, S3) through which Fl acts on Sl.

If a second substance cannot be introduced use standard 10.

8. Two substances interact but one or both substances or their interaction is badly susceptible to control (inspection, measurement, change); replacing them by others is ruled out; effective control is required.

a. Introduction of a field (predominantly electrical, magnetic or optical), passing through the system and "extracting" information on its condition.

b. Introduction of field F acting differently on Sl and S2 or acting only on one of these substances.

c. Constructing an S-Field with a battery (S2S3); the field F acts on S3.

9. The field and the substance interact but one or both of these elements or their interaction are only poorly subject to control (inspection, measurement, change); replacing the elements is ruled out; effective control is required.

a. Introduction of S2 interacting with F and Sl.

b. S2 is transposed into F' and F", an easily controllable S-Field is created of the elements Sl, S2 and F".

10. Two substances (or a substance and field) interact; one of them can be changed; the requirement is to establish (or improve) the second (supplementary) interaction (or action) without worsening the first (existing one).

a.   Constructing an S-Field, ensuring a second interaction in which the added field should not effect the first interaction.

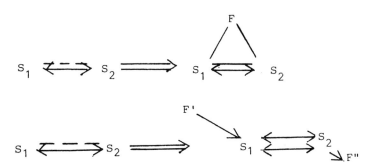

b. Construction of chained S-Fields, for instance:

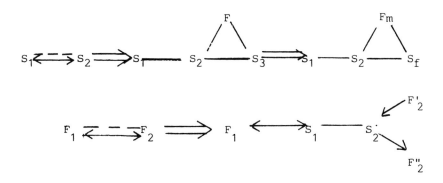

11.   Field and substance are connected by two conflicting attendant interactions; one interaction has to be removed, while preserving the other.

The introduction of S2 through which the field acts on S1, while this second substance is a part of S1 or a variation on S1:

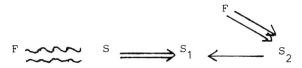

S1 "allows through" this action but detains the other.

12. Two substances interact; this interaction has to be
eliminated.
Standard 3: introduce a third substance, which is a
variant lof one of the given substances.

Type 3. Three elements are given

13. An S-Field is given which is only badly susceptible to
inspection or measurement; replacing and changing the
given S-Field is ruled out; the requirement is to ensure
effective inspection or measurement.
The S-Field given in the specifications of the problem is
regarded as a complex substance S2 (the problem has
virtually been transposed to class 1); a field is
introduced, for example:

If the S-Field contains a ferromagnetic substance it is
advantageous to introduce a magnetic field.

14. The same as in class 13 but one can replace or change
S2 part of the given S-Field.
The substance S2 is converted into an S-Field by the
introduction of S3 and F. A chained S-Field is formed:

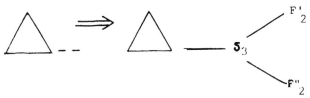

15. An S-field is given which is only poorly susceptible
to control; one can replace S2 and field F; effective
control is required.

The S-field given in the specifications of the problem is
reconstituted into a F-Field. The problem is virtually
translated into class 1: S2 and F are discarded, only
element S1 remains, which is augmented to make a full
S-Field by the introduction of a ferromagnetic substance
and a magnetic field.

16. The substance interacts well with field F1 but badly
with field F2; introducing new substances and fields are
ruled out; good interaction of S and F2 is essential while
preserving the interaction of S and F1.

The substance S1 subdivides into S'1 and S"1. The field F1
acts on S'1, the field F2 on S"1. If these actions are
incompatible in time, S1 subdivides in such a way that
they become in turns first S'1 and then S"1 and one action
is carried out in the pauses left by the other (standard
7).

17. The field F1 reacts very well with substance S1 but
badly with substance S2; introducing new substances and
fields is impossible; effective interaction of F1 and S2
while conserving the interaction of F1 and S1 is
essential.

a.   Field   Fl   subdivides into F'l and F"l. Field F'l acts on   Sl   while   F"l   acts   on   S2.   If   these   actions   are incompatible   in   time,   Fl is subdivided in such a way that it   becomes   in   turn first F'l and then F"l and one action is carried out in the intervals of the other.

b.   Field   F"l   is introduced which is identical in nature with   field   F'l   but   opposed   to   it   in   direction   (the "anti-pole"):

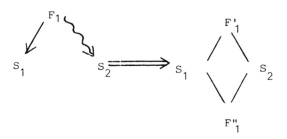

18. An S-field is given which should be eliminated.
This   problem   is   converted   to   class   12   and   solved   according to standard 3.

APPENDIX 3

THE APPLICATION OF CERTAIN PHYSICAL EFFECTS AND PHENOMENA
IN THE SOLUTION OF INVENTIVE PROBLEMS

Required effect, property	Physical phenomenon effect, factor, method
1. Measuring temperature	Heat distribution and the change it causes in the inherent frequency of vibration. Thermoelectrical phenomena. Spectrum of radiation. Changes in optical, electrical, magnetic properties of substances. Move through the Curie point. Hopkins and Barkhausen effects.
2. Lowering temperature	Phase transitions. Jowlie-Tomson effect. Rank effect. Magnetic calory effect. Thermoelectric phenomena
3. Raising temperature	Electromagnetic induction. Vortical currents. Surface effect. Dielectrical heating. Electronic heating. Electrical charges. Absorbtion of radiation by the substance. Thermoelectrical phenomena.
4. Stabilizing temperature	Phase transitions (including the move through the Curie point)
5. Indication of position and location of object	Introduction of marker substances; transforming the external fields (luminescent traces) or creating their own fields (ferromagnetic) and hence easily inspected. Reflection and emission of light. Photo effect. Deformation. X-ray and radioactive radiation. Luminescence. Change in electrical and magnetic fields. Electrical discharges. Doppler effect.

Required effect, property	Physical phenomenon effect, factor, method
6. Controlling location of objects	Action of magnetic field on object or on ferromagnet linked to the object. Action of electrical field on charged object. Transfer of pressure of liquids and gases. Mechanical oscillations. Centrifugal forces. Heat distribution. Light pressure.
7 Control of movement	Capillary action. Osmosis. Toms effect. Bernulli effect. Wave movement. Centrifugal forces. Weissenberg effect.
8. Control of aerosol flows (dust, fog, smoke)	Electrisation. Electrical and magnetic fields; Light pressure.
9. Forming mixtures	Ultrasonics. Cavitation. Diffusion. Electrical fields. Magnetic field in conjunction with ferromagnetic substance. Electrophoresis. Solubilization.
10. Separation of mixtures	Electric and magnetic separation. Changing apparent viscosity of separator liquid under effect of electrical and magnetic fields. Centrifugal forces. Sorbtion. Diffusion. Osmosis.
11. Stabilization of position of object	Electrical and magnetic fields. Fixing in liquids which harden in magnetic and electrical fields. Hydroscopic effect. Reactive movement.
12. Action of forces. Control. Creation of high pressures	Effect of a magnetic field via ferromagnetic substance. Phase transitions. Heat distribution. Centrifugal forces. Changing the hydrostatic forces by changes in the apparent viscosity of

Required effect, property	Physical phenomenon effect, factor, method
	magnetic or electric conducting liquids in a magnetic fields. Use of explosives. Electrohydraulic effect. Optical hydraulic effect. Osmosis.
13. Changes in friction	Johnson-Rabeck effect. Action of radiation. Kragelsky phenomenon. Oscillation.
14. Destruction of object	Electrical discharges. Electro-hydraulic effect. Resonance. Ultrasonics. Cavitation. Induced radiation.
15. Accumulation of mechanical and heat energy	Elastic deformations. Hydroscopic effect. Phase transitions.
16. Transfer of energy	Deformations. Oscillations. Alexandrov Effect. Wave movement including shock waves. Radiation. Heat conductivity. Convection. Phenomenon of reflection of light (light carriers). Induced radiation. Electromagnetic induction. Superconductivity.
17. Setting up inter-action of mobile (exchangeable) and immobile (fixed) objects	Use of electromagnetic fields (transition from "substance" to "field")
18. Measuring dimensions of object	Measuring inherent frequency of oscillation. Applying and reading magnetic and electrical markers.
19. Changing the dimensions of objects	Heat distribution. Deformation. Magnetic electrostrication. Piezoelectrical effect.
20. Checking of state and properties of surfaces	Electrical discharges. Reflection of light. Electronic emissions. Moire effect. Radiation.

Required effect, property	Physical phenomenon effect, factor, method
21.Measuring surface properties	Friction. Adsorbtion. Diffusion. Bauschinger effect. Electrical discharges. Mechanical and acoustic oscillations. Ultraviolet radiation.
22.Inspection of state and properties in volume	Introduction of "marker" substances transforming the external fields (luminescent traces) or creating their own fields (ferromagnetic), dependent on the state and properties of the substance under study. Changing the mean electrical resistance depending on the structure and properties of the object. Interaction with light. Electric and magnetic optical phenomena. Polarized light. X-ray and radioactive radiation. Electronic paramagnetic and nuclear magnetic resonance. Magnetic resilient effect. Move through the Curie point. Hopkins and Barkhausen effects. Measuring the inherent frequency of oscillation of an object. Ultrasonics, the Moessbauer effect. The Hall effect
23.Changing the volume properties of an object	Changing the properties of liquids (apparent viscosity, fluidity) under the action of electrical and magnetic fields. Introduction of a ferromagnetic substance and the action of a magnetic field. Heat action. Phase transitions. Ionisation under the effect of an electrical field. Ultraviolet, X-ray, radioactive radiation. Deformation. Diffusion. Electrical and magnetic fields. Bauschinger

Required effect, property	Physical phenomenon effect, factor, method
	effect. Thermoelectrical, thermomagnetic and magnetic- optical effects. Caviation. Photochomatic effect. Internal photo effect.
24. Creating a given structure. Stabilization of structure of an object	Interference waves. Standing waves. Moire effect. Magnetic waves. Phase transitions. Mechanical and acoustic oscillations. Cavitation.
25. Indications of electrical and magnetic fields	Osmosis. Electrization of bodies. Electrical discharges. Piezo and segneto electrical effects. Electrets. Electronic emissions. Electro-optical phenomena. Hopkins and Barkhausen effect. Hall effect. Nuclear magnetic resonance. Gyromagnetic and magnetic optical phenomena.
26. Indications of radiation	Optical acoustic effect. Heat distribution. Photoeffect. Luminescence. Photoplastic effect.
27. Generation of electromagnetic radiation	Josephson effect. Induced radiation. Tunnel effect. Luminescence. Hann effect. Cherenkov effect.
28. Control of electromagnetic fields	Screening. Changing state of environment, for instance, increasing or decreasing its electric conductivity. Changing the form of the surface of bodies, interacting with fields.
29. Controlling light. Light modulation	Refraction and reflection of light. Electrical and magnetic optical phenomena. Photoelasticity. The Kerr and Faraday effects. The Hann effect. The Franz-Keldysh effect.

Required effect, property	Physical phenomenon effect factor, method
30. Initiation and intensification of chemical changes	Ultrasonics. Cavitation. Ultraviolet, X-ray, radioactive radiation. Electrical discharges. Shock waves. Mycellarian catalysis.

BIBLIOGRAPHY

1. Poya D. _Problem Solving_: Translated from the English. Edited by Yu. M. Gayduk, Moscow, Uchpedgiz, 1959
2. Poya D. _Mathematical Discovery_: Trans. Ed. I.M. Yaglom. 2nd Edition, Moscow, Nauka, 1976
3. Adamar G. _Research into the psychology of the process invention in the field of mathematics_. Moscow, Sov. Radio, 1970.
4. Engelmeyer P.K. _Theory of Creativity_, SPb, 1910.
5. _Psychology of Thought_, Collection of translations: Ed. A.M.Matyushkin, Moscow, Progress, 1965.
6. Linkova N.P. _Use of ASIP as methodology for the study of the inventor's activity_. Included in book: _Problemy metodologii projektowania_, Warsaw, PWN, 1977.
7. Lapshin I.I. _Philosophy of invention and invention in philosophy_, Petrograd, Nauka i shkola, 1922.
8. Rainer M. _Rheology_, Transl. from English. Ed. E.I. Grigoliuk, Moscow, Nauka, 1965.
9. Dixon J. _Projecting systems: invention, analysis and acceptance of solutions_, Transl. from English. Moscow, Mir, 1969.
10. Altshuller G.S., Shapiro P.B _On the Psychology of inventive creativity_. Voprosy psikhologii, 1956, No. 6, pp 37-49
11. Altshuller G.S. _How to learn to invent_, Tambov Book Publ. 1961.
12. Altshuller G.S. _Foundations of invention_. Voronezh: Central Black Earth Publ. House, 1964.
13. Altshuller G.S. . _Algorithm of invention_, Moscow, Moscow Worker, 2nd Edit. 1973.
14. Altshuller G.S. . _Collection of problems and exercises on the methodology of invention_.
15. Altshuller G.S. _Basic methods of eliminating technical contradictions in solving inventive problems_, Baku, Gyandzhlik, 1971.
16. Altshuller G.S. _Analysis of solutions of inventive problems_. Incl. in book, _Materials for the seminar on the methodology of invention_. Inst. of Heat and Mass Exchange, Academy of Sciences of Belorussian SSR, Minsk, 1971, pp 51-133.
17. Altszuller G.S. _O uzdolnieniach wynalazczych_, Warsaw, Prakseologia, 1972, No. 41, pp 121-144
18. Altszuller G.S. _O teorii rozwiazywania zadan wynalazczych_, Warsaw, Prakseologia, 1977, Nos 1-2, pp 485-495.

19. Selyutskii A.B, Slugin G.I. Inspiration on demand, Petrozavodsk, Karelia, 1977.
20. Tsonev M.G. Methods types in eliminating technical contradictions in inventive creativity. Sofia: Central Soviet for Organization of Technical Creativity, 1977.
21. Voronkov V.D. Organising-Engineer's Handbook, Moscow, Moscow Worker, 1973.
22. Gutkin L.S. Basic trends in the theory of designing radio systems, Incl. in book: Radiotekhnika, 1976, vol. 31, No. 1, pp2-6.
23. Methods in the search for new technical solutions. Ed. A.I.Polovinkin. Yoshkar-Ola, Mari Book Publishers, 1976
24. Rubin I.D. Some paths in the development of electromagnetic consumption metres. Izv. vuzov SSSR. Oil and Gas, 1977, No. 5, pp 83-86
25. "Attention: Algorithm of Invention", Ekonomicheskaya gazeta, 1965, Sept 1.
26. Jones J.K. Engineering and Artistic Design: Transl. from Engl. Ed. V.F. Vendy and V.M.Munipov, Moscow, Mir, 1976.
27. Bush G. Methodological foundations of scientific control of invention, Riga: Liesma, 1974.
28. Deryagin R.I. Algorithm of the solution of inventive problems, Incl. in the book, Information Science and its problems, Ed. 5, Novosibirsk, Nauka, 1972.
29. Foundations of theory and general methods of patent evaluation, G.N. Anisov, I.I. Kichkin, N.M.Madatov, E.P.Skornyakov; Ed. V.N.Bakastov, Moscow, TsNIIPI, 1973.
30. Chinnov Yu.V. From my experience of solving inventive problems. Incl. in the book: Materials for the seminar on the methodology of invention, Inst of Heat and Mass Exchange, Acad of Sciences of the Belorussian SSR, Minsk, 1971, pp. 27-49.
31. Bogach V.A. We train inventors. Young Communist, 1972, No. 8, pp 85-88.
32. Kalenik V.I. Diplomas for inventors, Evening Dnepr, 1976, June 1.
33. Shuvalov V.N., Nasedkin A.I., Kulikov A.Yu. The "Golden Key" for seekers. Incl in the book: Economics and organization of industrial production, 1977, No.3, pp 193-197.
34. Chinnov Yu.V. The logic of success. Socialist Industry, 1972, Jan. 26.
35. Golovchenko G.G. The wind energy of plants. Physiology of Plants, 1974, vol. 21, ed. 4, pp 861-863.
36. Mitrofanov V.V., Sokolov V.I. The nature of the Russell effect. Physics of Solid Bodies, 1974, vol.16, No 8, pp 24-35.

INDEX

(continued from front of volume)

Other volumes in preparation

ISSN: 0275-5807